# BIG
# VEG

# BIG VEG

## GERALD STRATFORD

MY TIPS FOR GROWING VEG TO BE PROUD OF

HEADLINE

# Contents

**If you'd told me a couple of years ago that, at seventy-two, I'd be publishing my own book about gardening, I would never have believed you.** But the last year or so I have done many things that I never would have dreamed of. I've been interviewed for newspapers, TV, podcasts and radio and even been in a Gucci fashion shoot! And now I am lucky enough to be in the position to write a book. That's all thanks to the veg-growing community on Twitter. I am very grateful for all the interest my wonderful followers have taken in my gardening.

I wanted to write this book to provide encouragement for people who are interested in learning more about growing veg and especially big veg. I believe that it always helps to have a teacher for guidance when learning a new skill. I learned a lot from my dad, who was the most brilliant teacher I've ever had. While I wouldn't consider myself an expert, I have years of experience growing my own vegetables. Over that time, I've learned lots of lessons, and have tried to keep expanding my gardening knowledge. This book is my way of sharing that knowledge and experience. I'll tell you all about how I do things around my garden and what works for me, including tips for getting better results.

While I'm here, I'd like to take the opportunity to say thank you for picking up my book. Hopefully the following

pages will provide you some motivation to get out in the garden! Whether you're thinking about growing your own big veg, or just want to find out about the way I do things, I hope this book will be interesting enough to encourage you to get outside and have a go.

When you learn a new skill as a child, you can quickly take it on board. But when you get into adulthood, it seems to become harder. I used to teach fly-fishing and I noticed that youngsters would pick it up very quickly. But when I taught the adults, they would ask a million questions, become easily frustrated, and put other obstacles in the way of learning. My aim with this book is to remove any obstacles for learning. I find that I learn best when things are explained to me in simple, understandable terms. I hope I've achieved that and you'll find my book easy to understand. I have tried to explain words you might not know as I go along.

Before I get into the nitty-gritty, I'd like to share a little bit of background about who I am, and how I ended up here. I've packed a fair few things into my seventy-odd years. I've travelled to lots of different countries, had quite a few different jobs and made lots of friends. But gardening is something that has been with me for my whole life. Let me tell you a little bit about it.

> I hope this book can be interesting enough to encourage you to get outside and have a go.

## Early Memories

I was born in 1948, in a farmworker's cottage in the little hamlet of Worton, Oxfordshire. I don't actually remember the event (I suppose it was a little while ago . . .) but I've been told my grandmother helped as Mum gave birth to me.

The cottage was quite big, but it was very old. We had a large inglenook fireplace that I could almost walk into. There was only a cold-water tap in the kitchen. We didn't have any hot water at all so to have a bath, we had to boil the kettle on the range and fill up the tub. The toilet was outside. But that's how things were, back then.

I am the youngest of six children. I had two sisters, and three brothers: Rosebud, Nancy, Ralph, Brian and Arthur. My dad had worked as a farmworker in his earlier life, but throughout my childhood he was quite poorly with a lung condition. When Dad was ill Nancy became a second mother, cooking and looking after us.

On Sunday afternoons, one of our neighbours would gather us kids up and take us on walks across the surrounding fields. In the spring, we would pick wildflowers, primroses, bluebells and, my favourites, violets. I can smell them now. I would take them home for my mum – what an angel I was!

In the summer, everyone in the village would play cricket on the meadow next to the cottages. Boys and girls, and mums and dads. We loved cricket.

On a Sunday evening, we would walk across the fields to a little village called Begbroke and a pub called The Rising Sun that always smelt of polish and beer. Mum and Dad would go inside and have a couple of drinks, and one of them would bring us each a bottle of Coca-Cola and a bag of Smith's crisps. The bag would have a little packet of salt inside and you would sprinkle the salt on the crisps. That was a special treat.

In the autumn, we would pick blackberries and Mum would make blackberry-and-apple pie for pudding.

And then in the winter, my siblings and I would go out and collect wood for the fire. We would never cut trees down; we just collected dead wood that had fallen to the ground. I was sometimes trusted to use the axe to chop the wood when we got home.

I can remember, when I was getting adventurous at age twelve, I wanted to make something with the bits of wood I'd collected. So, Dad gave me a hammer, nails and a saw. I went in the kitchen and grabbed a chair, put the piece of wood on it and held it down with my foot. I was sawing away and thought, 'Crikey, this is taking a long time.'

**In the autumn, we would pick blackberries and Mum would make blackberry-and-apple pie for pudding.**

My mother came into the room and, by the look on her face, I knew I had done something bad. I let go of the saw, but it was stuck in the wood. I'd sawn straight through the chair. Mum kept that chair for years afterwards. I think it served as a reminder to be careful with the tools!

I might have made it all sound quite idyllic, and it was a very happy childhood, but life was harder, back then. My mother once told me that she didn't know why people went on about 'the good old days'. To her, and lots of people, they weren't that good. There was a lot of illness about, and food was hard to get. I was born just a few years after the Second World War and there was still rationing. Even when it stopped, my parents continued to live their lives frugally, and the mind-set stayed with my mother until she was ninety-three. Nowadays there is so much food waste, but back then we just didn't waste anything and that has stuck with me. I'll always try and find a way to use things, before I throw them away.

Dad and our neighbour used to have a pig each, and there was a pigsty at the end of the garden, where we fed them all our leftovers. The pig was our recycling bin!

Even though Dad was often poorly, that didn't stop him gardening, with the help of my three elder brothers, and me doing what I could.

First there was Arthur, who for all his life was known simply as Rebel. Dad had called him this when he disappeared as a toddler one summer's afternoon. The whole neighbourhood was out looking for him when Dad suddenly spotted something among the runner beans. He went to investigate, and there was little Arthur, fast asleep. Apparently, Dad said something like, 'There you are, you little Rebel,' and it stuck until the day he died. The next brother down is Brian, who I'm glad to say is still going strong. Then there's Ralph, who unfortunately died in 2010. Then comes me, the baby of the family.

Once I was old enough to join in with the others, I would try and 'help' in the garden. I probably just got in the way, but gardening is something I've done since I was tiny. I've had a fair bit of practice, since then.

I must have been about four years old when my dad took me out in the garden and showed me a bit of soil about the size of a large dinner table. He pointed to it and said, 'This is your garden, son.'

Dad gave me a packet of radish seed and left me to it. I'd watched him enough to know what to do. I sowed those seeds, and every day I would run down the garden to see if they were growing. Then, one day, I saw the first bit of green showing. From that moment, I was hooked!

## My favourite radishes

**Bacchus** – A round one that grows fast.

**Black Spanish** – This one has a black skin and white flesh.

**Candela di Fuoca** – A long, thin tapering Italian variety.

**Cherry Belle** – A mild, red radish.

**French Breakfast** – This is an easy one for beginners.

**Gaudry** – This one has a peppery taste and a real crunch.

**Mooli** – A large white one that cooks well.

By eight years old, I could dig alongside my brothers; Dad was in charge and we did what we were told to. Dad was a perfectionist. Sometimes we'd be planting things out and he'd ask us to pull them all up and start again. He never shouted or got angry, though. He'd just say things with authority, and we'd do as he said. Always.

When I was ten, Dad got a greenhouse. It was a second-hand one, made of red cedar. I remember him carefully cleaning around all the edges where the glass would go, then giving everything a coat of preserver before re-glazing the whole thing. He couldn't work fast because of his lungs, but he was very methodical. He used to spend hours pottering in his greenhouse.

By this time Dad was too ill to do any farm work. Our cottage in Worton had been 'tied', which means it came with the job, and so we had had to give it up. The council re-homed us just down the road in Cassington. After we moved, Dad took on three allotments in the village. It was a lot of work, but as we had eight mouths to feed and not much money, everybody in the family pitched in.

Dad was my idol and he was a great gardener. When I wasn't at school or playing with my mates, I would be with him, watching and learning. I suppose every young boy feels the same about their dad.

## Growing Up

I was sixteen, working in the market, with all the office girls around, smiling at me. It was quite a change from village life!

When I left school, I decided that I wanted to be a butcher, so Mum managed to get me an apprenticeship with a well-known butcher who had a shop in the famous Covered Market in Oxford.

Mum and Dad bought an old bike and did it up for me so I could cycle the six miles into Oxford to work. To me Oxford was the centre of the universe. I'd only been as a child with Mum before, I didn't have a clue what it would be like. The first thing that hit me about the city was all the beautiful women! I was sixteen, working in the market, with all the office girls around, smiling at me. It was quite a change from village life!

I soon became friends with lots of the other apprentices. In the market there were other butchers, fishmongers and fruit-and-veg shops, which all had apprentices. Within a week or two, the city on a Saturday night beckoned. We'd go to the Town Hall, where there would always be live music, or to the dance hall, Brett's, on Broad Street.

There were always a lot of Americans out in the city, because of the airbases nearby. They all had lots of stuff that we didn't. If you wanted stockings or other presents for your girlfriend, they seemed to have unlimited supplies of everything, from the base. For a lad from a little country village, this was all very exciting stuff.

For the first year or so of my apprenticeship, I was a glorified delivery boy. Being out and about meant it didn't take long for me to get to know people at all the colleges of Oxford's famous university. From them I would find out when the gigs were on. I saw The Beatles, The Rolling Stones, and many other great bands. Of course, at the time, I didn't realise how big these bands would go on to be.

I spent five years as an apprentice and three more as an improver. Working in Oxford in the sixties was a wonderful experience for a country boy. I met lots of wonderful people, and there was always so much going on.

One of my favourite memories is going to a summer ball at one of the university colleges. We turned up at 8 p.m. and it went on until 8 a.m. the next morning when we had our breakfast before going home. There were loads of things going on all night – it wasn't just about getting drunk and listening to music. There was lots of great food, and conversations, laughs, discussions . . . and girls! But less about that – you're here for the vegetables.

At the same time that I was working at the butcher's I was still supporting my parents back at home, fishing and gardening for Dad who was gradually getting sicker.

> I was still supporting my parents back at home, fishing and gardening for Dad, who was gradually getting sicker.

Unfortunately, Dad died in the November of 1967. We knew it was coming, but naturally the news devastated our whole family. My mum was one of the strongest women I have ever known. She did her grieving in private. On the evening after Dad died, Mum looked at us and said, 'What are you doing? Don't you have to be somewhere?'

By that, she meant the pub. At that time, the pub was the centre of our village life. When we said that we were staying to keep her company, her reply was, 'What for? I've got to get used to being on my own, so I might as well start now. Go on, get out of here.'

After Dad died, we all pulled together to help Mum carry on with the garden and greenhouse. Rebel and I looked after one allotment, and if Mum wanted any veg, she could help herself.

In the garden she grew dahlias, chrysanthemums and her favourite sweet peas. People would come from miles around to buy her flowers. It started off with the richer people in the village. They used to decorate the church with flowers. When they heard about Mum's wonderful flowers, they started coming to the house and buying from her. The word spread. On a Saturday morning, we would have buckets full of flowers ready to be sold by the back door. It was a way of making a few pennies for her.

> On a Saturday morning, we would have buckets full of flowers ready to be sold by the back door.

## The Cotswolds

I was blessed to find love again later in life with Elizabeth. We had been friends for many years as she had been married to my brother Ralph, and we grew closer after his death. One of the things we share is a passion for gardening. This is also shared by our families. Between us we have five children, nine grandchildren and five great-grandchildren so hopefully this love will pass down the generations.

Liz and I would both rather be outside than in. We are so lucky to live in a little bungalow with a large garden on the edge of the Cotswolds, near the picturesque town of Burford. We love it here.

When we arrived, just over four years ago, the garden was all just rough grass. After three months sorting and decorating the house, we made a start on the garden. First thing we did – with the help of Liz and Ralph's son Stephen and his son Bradley – was to build a large shed at the end of the garden. I call it my Cave. Then we erected a 12-by-8 greenhouse. Luckily, when I was tidying up, I found the concrete foundation of a previous greenhouse buried under about half a metre of rubbish, so, we put the new greenhouse on top of that. We also put up a fence between our neighbours and us. It's tall enough to keep

> **One of the things we share is a passion for gardening.**

Sky, our German Shepherd, in but still short enough to socialise over.

Once this was done, we marked out the area we wanted to keep as lawn and I dug the flowerbeds ready for Liz to take over. In this partnership, Liz is in charge of the flowers and seeds, and I'm the vegetable guy.

In the autumn of our first year in the bungalow, we heard that there were some allotments in the village that were vacant, so we asked if we could take a look. Historically there was a set size for allotments: 10 poles. It's an old measurement dating back to Anglo-Saxon times, apparently, and is the equivalent of 250 square metres, which is roughly the size of a tennis court. But these days they can be all shapes and sizes. We have two now – and two greenhouses and two sheds in the garden!

I'm one of those people who cannot do things by halves; I either commit to something or don't bother at all. Once we got to grips with the allotment, I started to think about growing a large pumpkin. I was talking about this on Facebook when someone suggested I make a Twitter account. Someone in the group told me how gardeners growing big veg use Twitter to chat with people about it. So I did. I started following a few people and some people followed me back, and that was that for a few months.

On 16 May 2020, during the first lockdown of the coronavirus pandemic, Liz and I decided to harvest a tub of early crop Rocket potatoes that I had been growing in the greenhouse. Rocket is a really early potato variety that we love to eat covered with butter with a salad. Yummy! So, I cut the foliage off the plant with some secateurs, put a piece of canvas on the lawn and emptied the tub on to it. I counted the potatoes out and weighed them with my fishing scale. It said something like 1.5 kg, not many, but for an early crop and brand new potatoes, they were fabulous. Liz had taken 'before' and 'after' photos so I just put a couple of pictures on Twitter with the caption: 'My first Rocket potatoes well pleased'.

Shortly after, my phone started tweeting and buzzing and making all sorts of noises. I thought it had broken. I didn't really know what was going on. So, I phoned up Stephen and asked him what was happening. I told him I'd posted a picture of my Rocket potatoes, and the phone had started buzzing non-stop. He said he'd look into it.

A little while later, Stephen called back.

'You've gone viral with your spuds!' he said.

I said, 'What's viral?'

Stephen explained that 'going viral' meant that people from all over the world were responding to my Tweet.

'You've gone viral with your spuds!'

Within seventy-two hours, I'd racked up something like 72,000 'likes' and 9,000 followers. I think I'd had about ninety followers before that.

The phone kept pinging and I didn't know how to stop it. I fell asleep with it on my bedside table but at 3 a.m., I woke up as it was still pinging. I had to get up and put it on the other side of the bungalow. (I don't think it's stopped buzzing and dinging since, but thankfully I've now figured out how to put it on silent mode!)

People were just enjoying these potatoes – something that's so simple and mundane in my eyes. I didn't really understand why it was happening. I think the lockdown must have been a factor; people have taken to gardening more over the last eighteen months. Perhaps my spuds gave people a moment to appreciate the simpler things in life. Who knows!

Of course, nothing has been the same since. I never would have dreamed of going on the TV or radio to talk about my gardening. Never in a million years would I have expected to write a book about my vegetables. But here we are. I'm very grateful to my lovely followers for all the interest they've shown in all I do. It's been a truly wonderful experience.

So, on to the veg . . .

> **Never in a million years would I have expected to write a book about my vegetables.**

**Whether you're already growing vegetables or you're just getting started, it's important to choose your veg wisely** – only grow what you can manage, and what you have space for. The last thing you want is to try and grow everything, only to end up neglecting your veg and having no end product.

It's hard for me to pick my favourite vegetables as I love so many different varieties! I'm lucky enough to have lots of space to grow almost everything I want to. However, there are some vegetables that I simply wouldn't go without, and I've listed these here for you. In the rest of this book I'll be describing how I grow some of these for showing, as well as sharing with you what I do month by month to get the best results.

One of the great joys of reaching thousands of people through my Twitter page is that people now send me their gardening questions. I spend four or five hours a day answering questions and I absolutely love it; if I can help somebody out and encourage them to grow their own, I always will. There are some things that I've been asked lots of times, so before we get started, here are my most commonly asked questions and my answers. Of course every gardener does things a bit differently, but this is what I have told my followers.

**Which seeds are best for a beginner veg grower?**

I always recommend radish, lettuce, mustard and cress seeds for beginners. Radishes are quick and easy to grow from seed, and can grow in the smallest of spaces. Lettuces are fairly low maintenance, and I always like that I can cut them and they come again!

Mustard seeds are so easy to grow, you just have to throw some seeds down on a bare patch of land, and give them a water – easy as that! Cress is even easier, you can simply lightly wet some kitchen towel and sprinkle the seed over.

**What would you recommend for people without much outdoor space?**

I would recommend starting with just one kind of veg. And make it something really useful and versatile like potatoes. You can grow potatoes in a bucket. All you need is a standard 10-litre bucket with some holes drilled in the bottom, and enough compost to fill it up. You start with a seed potato. When you're getting started you can buy these online, or maybe a gardener friend might give you one to try with. You need to plant it 10 cm from the bottom of the bucket, and you will get some lovely new potatoes.

A standard bucket filled with compost could also be home to some lovely carrots.

**The veg I just couldn't go without**

Potatoes

Carrots

Parsnips

Onions

Cabbages

Cauliflowers

Courgettes

Lettuces

Tomatoes

Cucumbers

**What would you grow if you only had a balcony?**

If you only have a balcony, don't worry! You can still grow a multitude of things.

You can grow cherry tomatoes in milk cartons and then hang them from the railings around the balcony. You need to start them off indoors, but once any danger of frost has passed you can move the little plants outside. To make the container, I cut a large hole in the top half of a 2-litre plastic milk carton on the opposite side to the handle, and make a few small holes in the bottom for drainage. Make as many containers as you need and plant a cherry tomato in each of them. That will give you plenty of tomatoes right through the summer. Liz and I love tomatoes, and I've told you all about how I grow them for showing, and how we eat them, on pages 58–65.

Alternatively, you could use the re-purposed milk cartons to grow radishes, which as I mentioned are very simple, cucumbers, lettuces, spring onions, sweet peppers or almost any small summer crops. In fact anything growing green in the vegetable line, if picked young enough, can be eaten as a salad.

You could even have your bucket of potatoes in the corner alongside containers with all your summer crops. The quirkier your containers, the better!

**TIP**

If you've only got a balcony, all you need is a trowel, some secateurs, gardening gloves and a watering can.

**Basic equipment**

Fork

Spade

Trowel

Rake

Push hoe

Secateurs

Garden line

Watering can

Gardening gloves

### Which fertiliser would you recommend?

If you have the space, use natural well-rotted compost. Another great fertiliser, if you can get it, is manure. I combine these with commercial fertilisers, such as fish, blood and bone meal, which is a slow-release compost, and growmore. Later in the book I've shared some of my own recipes to use for different veg. For tomatoes I always go for Tomorite.

### Would you recommend getting an allotment?

Allotments are great for those with no outdoor space of their own, or for people who have run out of room in their garden. I have my allotment, and I love it. But I'm a traditionalist, and I believe that if you have a garden you should definitely make use of that space before taking up an allotment.

### Are there any resources that you'd recommend?

I would recommend Twitter and other social media. There are groups that are especially for gardeners where you can ask a question, and someone will give you an answer – however silly you might think the question is. Like I said, I spend several hours every day answering questions on Twitter. Feel free to ask me on there if you're stuck!

**First and foremost, I am a gardener who simply loves growing vegetables to eat.** I love to share what I grow with my family and friends too. It's a great feeling, growing your own. But along the way I got interested in growing big vegetables to show in competitions.

About five years ago, I was reading one of Liz's gardening magazines. Inside, I saw a picture of champion veg grower Medwyn Williams showing some of his veg in preparation for the Chelsea Flower Show, one of the biggest events for gardeners in Britain. I remember he had these perfect onions and carrots – great big things – and it just set my imagination off. I thought that I'd quite like to have a go at growing something like that. I marvelled at people who could grow monster pumpkins and marrows, huge tomatoes or enormous onions. So, I started the following spring and I've been hooked ever since.

I do tend to buy a lot of my show and big veg seed from Medwyn as he's been growing veg for a long time and really knows his stuff. He's always been like a god of vegetables to me, and now I like to think that we are friends.

The showing season is from May to September. The real professionals work very hard to have out-of-season vegetables ready to show throughout those months. So at Chelsea and the other big shows, you can see everything from broad beans, which are one of the earliest harvests, to Brussels sprouts, which are a winter crop.

Most shows I have competed in have been held in village halls including the show in our own village, Milton-under-Wychwood, which is held in August. But I once went to a show held in a marquee in the garden of a stately home. The night before a show, I'm full of nervous energy. I'll be down in my Cave, pontificating, checking things over, and putting all my vegetables to bed. I might have a glass of cider or a can of lager, and if it's warm outside, I'll sit down with my veg and doze off with them. That might sound silly, but I love it. It helps calm the nerves. Both Liz and I will wake up very early on a show morning. In fact, it has been known for me to not make it to bed at all.

On show day, everything is loaded into the back of my Nissan X-Trail – not forgetting plates and other trays to display the veg on. I buy plastic disposable plates and spray them black for my displays of big veg. I think it helps the vegetables to stand out. New growbag trays are also useful for transporting and presenting my produce. I will enter different veg and varieties, depending on the different classes there are at each show and the time of year. At the last show I attended, back in 2019, I entered twenty types of vegetable. I had a truck full.

> Both Liz and I will wake up very early on a show morning. In fact, it has been known for me to not make it to bed at all.

Depending on where the show is, we set off to arrive as the venue opens for staging, which is usually at around 9 a.m. Staging means putting your produce out in the designated area. There will be a printed list with your name and all the classes you've entered. When you walk into the hall, everybody is staring and you never know what the competition will be like. It feels like a tense moment in a Western, except instead of pistols, everyone is clutching their veg. Then you have about an hour to put everything out.

I like to act very quickly. I want to get to the hall, get it all set up, make sure it's all looking good and leave. As soon as I see my spot, I will quickly create a mental picture of how I want my display to look. Then I just put the vegetables out, and go. I don't mess around! I don't want to hang about, because the more time you spend there, the more you're inclined to change something, lose something, fiddle around with things. I believe that most of the time your first choice is the best.

Sometimes there's a class for any vegetable that is not in any other class, so you can put anything in there. The only problem is that it's very difficult for the judge to decide between such varying vegetables. As well as individual categories, there might be a class where you display several different vegetables together – its usual to be given a

number, say four or five, of different veg to include. The idea is to make a pleasing display on a board and the judge chooses the one they think is the best.

I had a stressful moment at one show in 2019. I was entering the display class, but my carrots were too long for the table. When they're fresh, carrots are soft enough to bend, so I bent the root end round to fit them on! There wasn't enough room for the leeks, either, so I had to move things around, and squeeze everything on as best I could. I thought, 'Oh no!' But I suppose I must have been doing something right because out of twenty-one entrants in the show I had eleven firsts, six seconds, and one third. I was quite chuffed!

Once everything's set, you have to go away while the judges are in there alone. It's a very serious business! There are no names on the exhibits, so the judges don't know you from Adam. But you do need to label all your produce, because people – including the judges – like to have all the relevant information. Most shows will give vouchers or money as prizes. You also get a coloured card, a 'show ticket': red is for first place, blue is for second, and yellow is third.

Once we've dropped everything off, we have a few hours to kill. If it's a local show, we go home for breakfast or if we're too far from home, we take a picnic and find somewhere nice to spend the time. It's during this wait that the nerves kick in. But that's good, I think. If you're nervous, it shows you care. If you're not nervous and excited, find another hobby!

Once the judging is over, and we're allowed back in the hall, that's the big moment. When we walk back through the door, Liz checks one side of the table and I check the other. We will be looking to see what we've won, and it's such a lovely moment pointing things out to each other. Hopefully we'll go home with a few winners to our name, but it's a good laugh regardless. If it's a local show, a few of our family will come down and there's always lots of joking and laughing. You can't beat it, really; win or lose, it's a fun day out.

Now, I know I will never beat the professional giant veg growers, but the enthusiast inside me always wants to try. While I don't expect to ever get a world record (though I'll never say never) I'm happy to keep growing and improving. I always aim to beat my personal best. Even if it's only by one centimetre, it's still progress. I have my failures, like anyone else. The most important thing is to not allow yourself to become frustrated, and to learn from your mistakes.

Now where shall I begin . . .

> **Now, I know I will never beat the professional giant veg growers, but the enthusiast inside me always wants to try.**

**I love carrots. I love growing them, and I really love eating them.**

I start sowing them in the open ground at the allotment in March and finish in July to make sure we can enjoy them for as much of the year as possible. In October I'll dig up some carrots to store in sand. They can be left in the ground and dug as needed right through until Christmas, but if we want some when the ground is frozen it's handy to have some in the garage.

It's so easy to have a lot of something today and none tomorrow. It's better to plant a thin row of seed, and then sow again in two weeks to give you a regular supply and avoid getting lots of top foliage and no roots. With carrots, sowing thinly is extra important to avoid attracting carrot fly which can smell them from a long way away (see page 40).

The carrots I grow for showing are planted in barrels in the garden. They are just as sweet and tasty as any grown purely for eating. I compete in the Exhibition Long and the Stump Rooted classes.

Stump rooted carrots are shaped like a finger, but instead of tapering off, the carrot is blunt at the end. From sowing the seed to showing takes twenty to twenty-four weeks so if I'm aiming for a show at the back end of August. I sow the carrot seed by mid March.

# How I grow carrots to show

**Recommended varieties for competitions**

**Stump Rooted classes**
Sweet Candle

**Exhibition Long classes**
Long Own Reselection by Medwyn's of Anglesey

- I grow my long carrots in tall 200-litre barrels, and my Sweet Candle in the same size barrels but cut to half the height.
- Around February, I will give the barrels a good clean with a weak solution of bleach and drill some drainage holes in the bottom.
- Next, I put them on some sturdy shelving in the garden, about a metre off the ground. This is to deter the low-flying carrot fly. These dreaded pests lay their eggs on the carrot, then tiny larvae eat into the carrots and spoil them. For carrots in the ground, I put a barrier of fleece around the bed or a fence of fine netting. I also grow a pungent-smelling marigold called Tagetes in rows between those carrots. Liz can't stand it, but its scent stops the fly from smelling the carrots.
- But back to the barrels. I fill them up with builder's sand and give them a good soak with water. Then I leave them to settle.
- After they've sat for a week, it's time to bore out the planting holes in the sand. For this I use a 7.5-cm downpipe, cut to slightly longer than the height of the barrels, so I can reach the bottom comfortably when I'm making the holes. I drill two holes 10 cm below the top of the pipe and slide a wooden dowel through them both to make a handle.
- Using the pipe, I bore out six holes in the wet sand. I do this by using my strength to push the downpipe into the sand with a

backward and forward motion. When I can't go any further, I pull the pipe out and knock the sand out into a bucket. It usually takes at least two goes to reach the bottom. If you don't soak the sand beforehand, the dry sand would just refill the hole. The sand acts as a barrier to encourage the carrots to grow long and straight. In a barrel full of compost, the carrot would have more chance of growing in a bad shape, as there would be nutrients everywhere.

- I fill the holes with my carrot compost mix (see page 44). I made a funnel specifically for this job. I got three different sized flowerpots (15 cm, 10 cm and 7.5 cm) and cut their bottoms out. Then I taped them together with the bottom of one fitting into the top of the next one. It makes it easier to fill the holes without spreading the compost everywhere! I find it's best to do it with my hands rather than a trowel. I place handfuls of compost in the funnel until I think I am about halfway up. Then I get a cane and agitate the compost in the hole a bit to get rid of any air pockets. I then fill to the top and agitate it again. The first time I grew my carrots this way, I didn't think about air pockets. My carrot seedlings were growing happily and then one morning one of the carrots had disappeared 15 cm down into the barrel. The carrot was stuck and shouting for help. Poor little thing!

- To make sure that the water reaches the bottom of my tall barrels, I use a length of 5-cm plastic water pipe (again, a little longer than the barrel). Using a 5-mm drill bit I make holes up the length

**TIP**

When you're using a drill, always wear a pair of safety goggles and avoid baggy clothing. If you've never used a drill before, it's best to have someone more experienced guide you the first time.

of the pipe in a winding fashion. Then I find something to block the bottom of the pipe and push it down the middle of the barrel. I can now put a funnel in the pipe and pour water in that way. For this first time, soak the barrel well to settle the compost, then leave it for another five days.

- On 12 March I sow three seeds in a triangle in the centre of each tube of compost. This is to ensure that at least one germinates. I'm very specific with the days I like to do things, but any time in the middle of March would be fine. As I'm sowing them outdoors, I then cover each barrel with a sheet of glass to keep the temperature up a little. This also deters cats from doing their business amongst your precious seedlings!

- It takes 14–21 days for the carrots to germinate. The first true leaves will come after another week or two. When they have got their second true leaf, it's time to make a decision about which seedlings to keep. I examine the seedlings from every angle, I put my hand behind each one to assess them individually. I ask Liz to see which she thinks is the strongest, too. It's good to have someone else to blame if it goes wrong – even if I disagree and choose a different one. Once I have made the final decision, I take a pair of sharp scissors and cut the other carrot seedlings off level with the soil. I used to pull them up and plant them elsewhere, but it was pointed out by a grower I very much admire that this could disturb the roots of the one left behind.

## CARROT COMPOST

- 20 litres Levington Seed & Modular F2S

- 5 litres my own sieved, composted turf (see opposite)

- 285 g calcified seaweed

- 115 g Vitax Q4 plant food

- 85 g superphosphate

- 55 g NutriMate fertiliser

- 285 g fine vermiculite

*Sieve to mix and remove lumps.*

- When the seedlings are stronger, I put a couple of pieces of 2-by-2 wood across the opposite sides of the barrels and place the glass on top to add a bit of airflow around the plants. If there is a sharp frost forecast I take the wood out at night. By the end of April, the glass can be removed altogether as the carrots are hardy enough to go without it.

- Now it's just a case of checking if they need water each day. In April and May the long carrots can take up to 4.5 litres once a fortnight. In June, July and August I up this to once a week. The carrots don't need any fertiliser through the season; what I have in the compost is plenty for them. I usually say, 'Good Morning!' as I as I pass them to go about my other jobs – it's important to keep chatting away to them, I think!

- While the carrots are still quite small, I put a collar made from a piece of 7.5-cm pipe round the top of each one. Then if the carrot lifts out of the soil, I can just sprinkle some compost inside the collar. Otherwise the top would go green, which is a definite no-no in competition – it would be instantly disqualified.

- By the start of August, we're getting near to the witching hour. About two weeks before the actual show I like to lift one of each of my two sorts of carrots, to see how they're looking. This is a very exciting and scary moment. So much work has gone into them – what if they're a disaster?

- My first job is to give the barrels a good soak, so everything is nice and wet. This helps the carrots to come up with the all-important root complete. Every millimetre counts. So, I always make sure I am extra careful pulling them up.

- I gently scrape away some of the sand around the top then I put both hands around the top of the carrot and *very* gently pull and twist. This breaks the little root hairs along the length of the carrot, making it much easier to pull up. After that, a gentle lift, a release of pressure and up it comes . . . the moment we've all been waiting for!

- With any luck, the most beautiful carrot I've ever grown emerges and the smell is absolutely wonderful. I put it on my bench and wash it off very gently with clean water and a sponge. Never scrub at the skin. After all this, I look closely for any imperfections. Is it straight? Does it look even? Is the root intact? Hopefully it's as near perfect as I could hope for! But these are just the practice runs, so I still have to hope the ones I pull in two weeks' time will be just as good.

- On the evening before the show, I pull up all the carrots. I am looking for three long carrots and three stump root carrots that will be show winners. But who knows what they'll look like once they're pulled up? It's a bit like checking the lottery. Will I be a millionaire, or will I have failed to get the winning numbers?

- After lifting and washing them, I cut the leaves off, leaving 10 cm of stalks on the carrots. Then I lay the stump root carrots out side

## TIP

You can make compost from old bits of turf dug out of lawns. Stack them soil side up and leave them for up to a year to rot down, then put the compost through a sieve. I used the turf I dug up for the vegetable patch in our garden and from other people digging up their lawns.

by side, swapping them around until I get three matching carrots that are as perfect as I can get them. Three perfect smaller carrots will always beat three less perfect large carrots, but of course three large perfect carrots will beat anything! I go through the same procedure with my long carrots, only here I'm looking for length as well as everything else.

- Once I have got my carrots sorted, I tie the tops with raffia to keep the stalks nice and neat. Then I put wet kitchen towel on a plastic growbag tray and lay the carrots out. I cover them with more wet towel and put them in the cool of my Cave overnight, before the big day.

**Longest carrot**

6.245 metres
UK, 2016

**Heaviest carrot**

10.17 kg
USA, 2017

I was quite proud of these.

**I get so many people saying to me that parsnips are a difficult vegetable to grow.** This is because they are sowing their seeds too early! Parsnips are very sensitive to temperature, and if the soil is too cold, the seeds won't germinate. It's easy just to look at what it says on the seed packet, but if it's a cold spring the seeds are just going to rot in the ground. The best thing to do is to take the temperature of the soil, it needs to be about 12°C to sow outside. If it's lower then you can either wait for it to rise naturally, or you could put cloches over to warm the soil up quicker. Cloche is French for bell and traditionally these were bell-shaped translucent covers, but these days you can get them in all shapes and sizes. The cloches need to be kept in place until you can see the parsnips growing.

As well as my long ones for showing, I grow parsnips on the allotment. A good variety for growing outside is Tender and True, that's my favourite and the one I always have for us to eat. Roast parsnips, beautiful!

One thing to be aware of, which I learnt the hard way, is that parsnips have an irritant sap that's absorbed by the skin. When combined with sunlight, this can cause sunburn, itching and blistering. So if you need to get up close and personal with your parsnips, it's always worth wearing gloves and long sleeves!

## How I grow parsnips to show

- I grow parsnips in barrels, in more or less the same way I grow my carrots. There are a few differences, though. My parsnips can grow to over 1.5 metres long, so I put an extension on the top of the barrel to make sure they have room to reach their true potential. An extra millimetre could be the difference between winning and coming second. Also, instead of 7.5-cm holes, I bore out 15-cm holes in the sand. This means I can only fit four in a barrel. I follow the same routine as with the carrots but because the holes are bigger, I mix more of the compost. Once I have filled the holes they are left to settle.

- At the same time as preparing the barrels, I like to chit the parsnip seeds. This just means that I put some wet kitchen roll in a little plastic takeaway container, and sprinkle on as many seeds as I need, plus a few more in case of failures. Usually, I do around twenty. Most of those will turn out well, but you never know with seeds. You could have 100 per cent success one year, and the next year have 100 per cent failure. Once the seeds are on the wet kitchen towel, I put more damp tissue on the top, and put the lid on the container and leave the seeds to stand in a warm place.

- I check the seeds twice a day, and the minute I see a little white tip of root showing get them straight down to the heated greenhouse. I'm very careful when I move them, even in this very early phase. I make sure to hold them inside my coat, just to make sure they don't get chilled.

- In the greenhouse, I prepare a twenty-four-cell module with seed compost, filling each cell almost to the top. I buy my seed compost ready mixed from Medwyn's of Anglesey, but any seed compost from a reputable firm will be good. Then I use tweezers to place two seeds into each cell. I follow this with a layer of vermiculite over the seeds – you can use compost, but I just prefer vermiculite.

- These are grown in the greenhouse until March when, as long as the weather is not bitterly cold, they will be ready to transfer to the barrels. First, I fill the holes in the barrels with my parsnip compost mix, then I make a hole in the middle of the compost that is big enough for the whole cell to fit it. I gently remove each seedling in its compost from the plastic cell and plant it in the barrel. Finally, I put a sheet of glass on top of the barrels to protect the seedlings from the elements (and the cats!).

- I look after parsnips in the same way as I do the carrots, checking them each day, watering as needed and, if they need pepping up, spraying with the Epsom salts mixture (see page 42). It's quite easy to look at them and the carrots together and do any maintenance needed.

- Once they are growing well, I remove the smaller of the two plants in each hole by cutting it off next to the soil.

## PARSNIP COMPOST

- 75 litres Levington Seed & Modular F2S
- 25 litres my own sieved, composted turf
- 455 g calcified seaweed
- 140 g Vitax Q4 plant food
- 115 g superphosphate
- 85 g NutriMate fertiliser
- 285 g fine vermiculite

*Sieve to mix and remove lumps.*

- When it comes to harvesting parsnips, I follow the same instructions as for carrots, but with two differences. One is that before trying to pull the parsnip, I cut the leaves down to 15 cm. The second is that as the barrel is taller and the parsnip is longer, I use a small set of steps to get above the parsnip that I am going to pull.
- Once you've loosened the parsnip, make sure you pull it very carefully and slowly because that root is long and thin and you want every last millimetre on the end. After pulling them up, washing them, and choosing the three best examples to show, these too are put between damp paper on a growbag tray and left beside the carrots in the Cave.

**WORLD RECORD**

**Longest parsnip**
6.55 metres
UK, 2017

**Heaviest parsnip**
7.85 kg
UK, 2011

**Beetroot is another veg that you can eat all year round if you sow little and often.** The thing about beetroot though is that every 'seed' planted is really a group of seeds, so they come up in bunches. I thin them out to one every 15 cm or so.

It's a good idea to dig them up in October because they aren't fully frost hardy. When you lift them don't be tempted to cut the leaves, or the beetroot will 'bleed' and go pale – just twist the leaves off the top.

I then pickle them to store. I just boil them until they're cooked, then plunge in cold water. As soon as they are cool enough to handle, the skin can be rubbed off. When fully cold, I slice and pickle them. If they are small, they get pickled whole.

I also make a beetroot chutney that tastes to me like the beetroot salad you can buy in supermarkets, but it's cheaper to make and lasts longer. You need to boil, chop and cool 1 kg of beetroot. Then peel and chop 500 g of cooking apples and two medium onions, and boil these up in a maslin pan with 660 ml of malt vinegar and 200 g of demerara sugar for about 20 minutes, until the onions are soft and the mixture is fairly thick and golden brown. Remove from the heat and add the beetroot, allow to cool and transfer to sterilised jars.

## How I grow beetroot to show

**Recommended variety for competitions**

**Exhibition Long classes**
Reselected Extra Long Black Beet

- If you are growing beetroot to show, you can either grow them to get the longest root (they don't need to be fat), or you can enter them in a Globe class. I grow mine like a carrot and compete in the Exhibition Long class.
- For my beetroot, I use the same method and barrels as I do for carrots. I bore out five or six holes and fill them with the same compost mix (see page 42).
- I sow beets in March at the same time as the carrots – again, with three seeds in a triangle to each hole full of compost.
- The plants are nurtured in the same way as the parsnips and carrots, but you need to take a little more care not to over water as I think it can affect the shape. I want my beetroot to be a nice, long, cylindrical shape, tapering off easily – but sometimes they grow awkwardly and the beet looks twisted instead. You don't want that to happen: the judges in a show wouldn't like your beets to look untidy. I think irregular watering might have something to do with it; if you're giving a plant 1 litre of water today, then 2 litres tomorrow, you're not giving it the stable conditions it needs. If you water irregularly, the plant goes searching for water, which causes it to twist and turn in different directions as the roots search. So I always try to be consistent with how often I water my vegetables, and to make sure I'm

giving them the same amount every time. But that's not gospel. That's just how I see it!

- One thing to be aware of with beetroot, is that they are inclined to get a pest on the leaf called a leaf miner. Its larvae will cause damage by tunnelling into the leaves and feeding on their internal tissues. You'll know when they are there because the leaf starts to wilt, and it has little lines running through it. It's very hard to prevent, but I spray them with a tiny bit of pure, non-detergent soap mixed with water to mask the scent that these little nasties seek out.

- As with carrots, beetroots are also lifted on the evening before the show. Again the leaves are cut down to within 10 cm of the crown. Once I've selected the three I am going to show, these join the carrots and the parsnips in the cool of the Cave.

WORLD RECORD

**Longest beetroot**
8.56 metres
UK, 2020

**Heaviest beetroot**
23.995 kg
UK, 2019

**Liz and I love tomatoes, and the natural follow-on from that is growing them for show.**

We eat tomatoes every day. Can't get enough of them. If there are any left over that we don't manage to eat, Liz might make her curry tomato chutney. We'll also take this opportunity to use our little gadget the dehydrator. After about twelve hours on there, the tomatoes are all dried out. We put them in olive oil and – bingo! – sun-dried tomatoes. We also freeze tomatoes to use in slow cooks during the winter. I usually grow several varieties and only sow a few of each tomato at a time as I want to stretch the season out and have fresh tomatoes for as long as possible.

When it comes to showing, there are so many different classes for tomatoes: Standard, Plum, Cherry, and so many different colours. I've done well in the Standard class in the past.

With the monster tomatoes weight is everything. They are actually very ugly to look at – they grow in funky shapes, not symmetrical at all. But that doesn't matter – they are still very tasty to eat. As yet, I haven't entered a large tomato in a show. I grow them, but right now I'm just curious to see how big the tomato can grow! I was sent some seeds for a large one called Megadom this year so fingers crossed!

## How I grow tomatoes to show

- I grow all of my tomatoes straight into the soil. Tomatoes have a strong root system and they're a very hungry plant. In the ground there's a larger area for roots to explore so I find I get better and larger tomatoes this way.
- The first thing I do is start preparing the ground in the small greenhouse where the tomatoes will be spending most of their lives. In January or February, I dig out the borders either side of the pathway, to make a trench about 30 cm deep. Then I put a 15-cm layer of rotted horse manure in the bottom, then 10 cm of my own compost mixed with fish, blood, and bone meal and crushed oyster shells. Next is a sprinkling of chicken manure pellets covered by a 10 cm layer of my compost. Finally, I will give it a good watering and let it settle.
- I sow my first batch of seeds about 20 January. I sow three or four seeds in each 7.5-cm pot, cover with vermiculite, water, and put the pots in the propagator. The seeds start showing after four or five days and by 4 February they are ready to be planted up into separate 7.5-cm pots. I fill them with compost, cover with vermiculite and put them back in the propagator, making sure to label them. We sow again on 22 February, my main sowing date.
- About six weeks after the seeds are sown they can be moved from 7.5-cm to 11-cm pots.

- Each time they are re-potted, we make sure to bury the seedlings in the compost to just below the leaves because tomatoes have the ability to produce roots from all the little hairs on the stem This will also give them more stability.
- The week before I start to plant the tomatoes out in their final place, I place tomato 'haloes' about 40 cm apart. I do this in a staggered row as my borders are wide enough. A halo is a clever little invention. They're made of plastic, usually round, but can be oval shaped. There is a hole in the middle, surrounded by a double wall. The tomato is planted in the middle and water is put in the reservoir between the walls. The idea is to push the bottom into the soil, so it is a snug fit; you don't want water running out around the edge.
- Once the haloes are in place, I put a sturdy bamboo cane in the centre and tie them up to the support frame that I keep permanently in the greenhouse. You can use bamboo canes for that too, but I wanted something really sturdy, so I made a frame out of roofing battens (I find a lot of uses for these!).
- By the time I start planting out, I make sure to have the heating set to come on if the temperature drops below 10°C because otherwise the tomatoes wouldn't survive. I start with just a couple of the early-sown tomatoes in the middle of March and then plant out a couple at weekly intervals until the greenhouse is full. Any extras get planted in containers or grow bags around the garden.

## My favourite tomatoes

**Sungold** – A lovely sweet yellow cherry.

**Black Opal** – A very small dark purply black cherry.

**Sweet Million** – A cherry tomato that's ideal as a snack.

**Shirley** – Our all-time favourite, perfect for sandwiches.

**Y Ddraig Goch F1 (Welsh Dragon)** – A beautiful, uniform standard tomato.

**Gigantomo** – A beefsteak tomato that has a lovely flavour.

- As the tomatoes grow, I gradually fill the middle of the halo with more compost. This grows more roots to nourish the plant.
- When the first truss of tomatoes starts to flower, I go around with a small, soft artist's brush and hand pollinate them. I do this by gently brushing across each flower to spread the pollen around.
- When the first fruit shows, it's time to start feeding them. To start off, I feed once a week pouring most of the feed – I use Tomorite and just follow the instructions on the bottle – into the middle of the halo. As the tomatoes grow, I feed more often. Soon it will be twice a week, and then every other day. At the height of the season, they will be fed every day. Once a week I fill the halo reservoir with Tomorite solution. On the other days the feed goes in the middle and water in the reservoir.
- I spend time each day loosely tying them in to the canes and pinching out any side shoots. They need the tops pinched out once there are six trusses (bunches of fruit) so that all the energy can be directed into those. By now, the leaves on each plant need removing, up to the bottom truss. This helps with air flow and ripening of the fruit. Maintaining a good level of air flowing through the greenhouse is important at this time of year, because stagnant or damp conditions can encourage potato blight. This disease affects tomatoes as they are members of the same family. When my tomatoes are coming along in the greenhouse, I put the radio on for them. It keeps the bad vibes away!

**TIP**
I always water after I've sown seed or pricked out seedlings. A lot of people soak the compost first, but I find my way works for me. Make sure to water everything from below by standing the pots in a gravel tray full of water. This causes less disturbance – especially with seeds.

**Heaviest tomato**

4.896 kg
USA, 2020

- To grow a big tomato, the basics are the same but when the plant starts to flower, I am looking for something a bit different. Usually for tomatoes, it's one flower per fruit. But for these big tomatoes, you need to have a few flowers growing close together. If I'm lucky I will find a flower that is actually two, three, or even four flowers joined together. Then when they grow, they fuse together and grow into a big tomato.

- I pollinate this with my artist's brush and cut all the other flowers off that truss. If I'm really lucky, I'll find another super flower and do the same with that. Then I remove all the other trusses of flowers and just leave these two to grow on, praying they will be ready by show time. Tomatoes are quite a delicate fruit, and don't last at the peak of perfection for very long. But these things are out of our control. It's all up to fate! On the evening before the show, I delicately cut the best tomatoes and match them up to get the most perfect and similar group. It's important to check the show schedule to make sure how many tomatoes are required for the class you're entering. It varies from class to class. It might be five standard tomatoes or seven cherry tomatoes.

- I cut them off the vine carefully because I need to keep the calyx – that's the green bit at the top – intact. After choosing the tomatoes, I get a damp piece of tissue and gently wipe them over to remove any water marks on them. Then they're put on a tray and tucked away with the other veg for the night.

**Last year, we grew over a hundred mammoth onions.** Most people as they get older tend have less up the allotment, but I seem to have more and more!

We didn't really expect to have so many. We were originally planning to take some along to a boot sale, where people like to buy young plants as they're cheaper than from a garden centre (and I like to think mine are better plants), but of course it didn't happen because of Covid. So, instead I planted them up in the allotment. I think some people got a bit jealous, and it was a bit of a talking point! One woman who stopped by was amazed at how big one of the onions was. This one in question was less than a kilogram – not a monster, by any stretch. So I pulled it up there and then, and handed it over to her. I was just happy for her to go home and eat it!

All of my big onions are planted outside eventually. Most people who grow enormous onions have the benefit of a much larger greenhouse or a polytunnel. This allows them to grow their onions in large pots all the way through. I do have plans for a polytunnel when funds allow. Then I could grow even bigger onions. But at the moment I'm quite happy if each year I can beat my personal best, even if only by a few grams. Anyway, time to stop dreaming and get back to work!

## How I grow onions to show

- I sow my onions on Boxing Day. I would sow them on Christmas Day, but that's the one day of the year that I'm barred from working in the garden! One day's holiday a year can't be bad. I sprinkle the seed fairly thinly on a seed tray filled with compost and cover them with a thin layer of vermiculite. Then I gently press down with my hand just to be sure the seed is in contact with the soil. Once I've done that, the seed tray's put in a gravel tray of water to draw up moisture from the bottom. I'm convinced watering this way helps to stop damping off, a fungal infection that causes the seedlings to collapse and die.
- They are in the electric propagator until they pop up. This takes around a week. Once the seeds appear, I move the trays to the heated bed in the large greenhouse, without any cover over them.
- About a week later they should be ready to prick out. The best time to do this is while they are still in the 'hook' stage. This is when the stem is doubled over, with the seed still in the soil or just on the surface. Over time, the stem straightens up so Liz and I do try to plant them in modules before that happens. As with everything else, I water them from the bottom after potting up.
- We use the twenty-four-cell modules in a standard seed tray for everything we transplant. I fill the cells with seed compost, and make a hole in the middle in each cell. I then scatter some

mycorrhizal fungi powder in the hole, before planting the seedling and firming it in. Mycorrhizal is a naturally occurring fungus in any healthy soil. It has a symbiotic relationship with plants; the fungus uses the plant's sugar and carbon reserves as energy and in return the fungi supply water and nutrients that the plant needs. It's almost like having an extra root system, so it really helps to give the plants a boost.

- Once transplanted, the seedlings are put back on the heated bed for another week to get going before I move them to the other side of the greenhouse away from the bottom heat. If they were left on the heat, the tops would grow faster than the roots can support, and they would be spindly and weak. The greenhouse is still heated of course. It doesn't go below 15°C at night to make sure we keep everything ticking over.

- At the beginning of February, the onions are potted into 7-cm square pots. Each onion is given a support to help keep its leaves upright. This improves growth, because the minute the leaves start flopping, they are not really doing much and could be damaged. You can buy supports that are purpose-built for this. But they're simple enough to make yourself.

- When the plants are small, my supports are made with bamboo skewers (the sort you use to grill a kebab!) and a length of foam cut from a 5-cm lagging pipe. This pipe has a long slit running from top to bottom that is perfect for sliding the leaves inside.

**TIP**
I always use tap water for seeds and seedlings. I carry a couple of cans into the heated greenhouse twenty-four hours before I need it. This allows the water to reach an ambient temperature and the chlorine to evaporate. Never water your seeds with rainwater as it can cause damping off, a disease that makes seedlings collapse.

5-cm lagging pipe

Dowel supports as the onion grows.

- I cut the pipe into pieces about 2 cm wide, then push the kebab stick down the side of the pipe opposite the split. Then I carefully stab the skewered pipe into the pot and gently slide the leaves through the split in the foam. Simple! It's easy to adjust the support as the onions grow by just sliding the lagging up the skewer. Later on, I will swap the skewer for some dowel.
- If everything goes well, the onions will move on to 11-cm pots by the end of February. Then, they are allowed to grow on until the end of March. At this point, I need to harden them off. Onions are classified as hardy, but no plant can cope with going straight from a heated greenhouse to a cold garden without suffering a drastic setback. Hardening them off involves putting them in a closed cold frame and hoping for some nice weather, allowing you to open the frame up a little each day (but closed at night) for a week.
- Then, as long as no hard frosts are forecast, I leave the cold frame roof open day and night for another week. Towards the end of April, if there's no sign of frost for a couple of nights, I plant them out in my onion bed. It's important to ease them in!
- The onion bed is in my garden, not at the allotment. I keep them close by so I can keep an eye on my babies! I use the same piece of ground every year. Onions are the only exception to the rule of rotating crops as the diseases that onions get can be controlled. The company that produces the seeds for my onions has been

## TIP

Keep an eye on the onions growing. If the bulb gets close to touching any of the supporting canes, move them away a bit. You don't want dents on your beautiful onions because they have started to grow around a cane or two!

using the same land for over a hundred years. If it's good enough for them, it's good enough for me. I just have to make sure that the land is well looked after. In the autumn I add a layer of well-rotted horse manure and dig the bed over. As my soil is so sandy, I cover the whole bed with strong black polythene sheeting to stop the rain washing the nutrients out of the soil during the winter.

- Two weeks before planting out I lift the polythene and check the pH levels in the bed. Onions don't like it too alkaline, they need a pH of 5.5 to 6.5. If the reading is high, I will add some sulphur chips to make it a bit more acidic. Otherwise, all I need to do is rake the soil to a fine tilth ready for planting.

- I stagger the planting. I'll plant two rows and a week later, another two, until I've planted them all out. To grow to their full potential, these onions need more room than normal onions. If they're planted too close together, they could also get mildew or botrytis, which kills the leaf and means the onion will be smaller – there's nothing worse than watching those precious onions struggle.

- First, I put a garden line across the plot and place the pots down every 45 cm. Then I dig the holes just big enough to fit the plant in. If the roots are tightly packed around the edge of the pot as I remove the plant, I gently tease them out. Otherwise, they could carry on growing in a tight circle and it will never really get going. Before I put the plant in the hole, I sprinkle mycorrhizal fungi powder round the plant roots and in the hole. Then I place the onion in the ground gently and firm the soil around it. Next, I move the line over 45 cm to make the next row.

- I place the next row of pots at 45 cm apart along the new line, with each plant mid-way between the plants I put in the first row. I do this so they are in a diamond pattern, which makes it easier to hoe the ground, and so allows more oxygen into the soil.

- As I plant, I replace the little kebab-stick supports with three or four 1-metre bamboo canes around each onion, making sure I don't touch the bulb with them. Then, with soft garden twine, I tie around the canes to help hold the leaves up. As they grow, the leaves will get very large and heavy and need this support.

- Once they are all planted, don't forget about them! I look at my onions every time I walk up the garden and every time I walk back down to the house. Is one of them looking sickly? Is that one going to seed? Is anything else not looking right? If something doesn't look as it should, I go and investigate.

- A sickly onion is pulled straight out and thrown in the bin. Don't chuck it on the compost heap. Whatever it's suffering from, you don't want it in your compost to spread back in the garden.

- If I see one starting to throw a flower shoot – known as bolting – I pull it out too. If it's going to flower and go to seed, it won't be putting strength into making a large onion. If it has some growth on it, we will eat it, usually sliced up on a salad.

**TIP**
You can make a garden line from strong string wound around two short stakes. Plant the onions on the far side of your garden line – away from the direction you're working in – so you don't damage the leaves when you move the line.

- I weed the bed regularly. Weeds compete for moisture and nutrients; you don't want to be sharing any of that around! I also keep an eye out for the odd one or two putting on more growth than the others. If I see this, I lavish extra care and attention on these onions as they might be the ones to beat my personal best. To me, that is far more important than any show ticket.

- Onions need to be lifted two weeks before the show date in August so they can ripen and look their best. To lift them, I borrow Liz's 'lady' fork – a border fork that is smaller than the full-sized fork used for heavy digging. I slide it into the ground a little way from the onion so there's no risk of damaging the precious bulb and gently push down on the handle to loosen the soil around the roots. I try to keep the roots as intact as I can. Then I put my hands underneath the bulb and ease it out.

- I take the onions to the table by my Cave where I do all my veg prep, and take a look at them. There are several different categories for onions. I might need a set of three or a set of five. I select those I think are the best. I could find I've got an embarrassment of riches or, heaven forbid, I might be struggling to find enough.

- Next, I give my selection a quick weigh, mainly to satisfy my curiosity, then they are laid on the staging in the large greenhouse. There is room in here because by this time of year we only have chillies, peppers and aubergines growing.

- After a couple of days, the leaves start to wilt so I cut them into a spear shape to tidy them up. I don't want dying foliage hiding something unwanted. I gently wash off any loose skin and turn them daily to help them dry and ripen.
- On the evening before the show, I cut the roots off with a pair of scissors, leaving about 5 mm. Then with a sharp knife, I cut more of the leaves off. The aim is to leave about 10 cm above the bulb, and I bind this with raffia to make it look neat. Lastly, I weigh them all as they can lose weight while they are drying out. I then make my final decision about which is going in each class.
- This is when I find out if I have beaten my personal best, which is 2.7216 kg. Not bad for an outside onion!

**Heaviest onion**
8.5 kg
UK, 2014

**Potatoes are one of my favourite vegetables to grow.** I think it's due to the excitement of not really knowing how they're going to turn out until they're harvested.

I think everyone should grow potatoes in school. If every child grew just one potato in a bucket, by the end of the summer term they would each have enough to take home a few potatoes for their families. I think it's so important for children to know where their food comes from, and how it's grown. It's also rewarding and I think it would get more people interested in growing.

When growing big potatoes, I don't try to grow the heaviest single potato. There are lots of people who enter competitions just for this, but that's for the real specialists and unless you're incredibly lucky, you'd have to indulge in buying an expensive potato called Condor.

My challenge is to see how heavy a crop I can get from four potatoes planted in a 30-litre tub. My personal best is just over 6 kg. Not bad, from four potatoes!

I grow up to fifteen varieties each year, with one or two tubs of each variety, plus a few more in the open ground to feed us through the winter. Potatoes can be all colours and this year I'm trying one called Salad Blue. Last year I grew eleven varieties, and they came out at an average of just under 4.5 kg per bucket. It's a good job I like spuds!

## How I grow potatoes for high yields

- Potatoes are grown from seed potatoes. These are potatoes that are stored over winter. This is called dormancy. If you don't have a period of dormancy, the potato will not re-grow and instead it will rot. I like to think it's Mother Nature at work – shutting down in the winter and saving all her energy for the following summer, like a tree losing its leaves.

- In the middle of January, I 'chit' my first and second early seed potatoes, followed by my maincrop a couple of weeks later, at the start of February. Chitting is the process of encouraging the potato to sprout before they are planted. I think we've probably all experienced getting an old potato out of the cupboard and finding it is covered with roots and shoots! Well, that is exactly what we want to happen as long as they're sprouting strong, little shoots, not long straggly ones.

- I store the potatoes in shallow wooden trays with handles that I make specifically for this task. They each hold two large egg trays. I place one potato in each space, with the side with the most eyes – the little indentations on the potato that sprout if left too long in the warm – facing upwards. Then it's as simple as leaving the trays in the daylight in my Cave, just checking them once or twice a week to make sure none have gone bad. As long as they get a bit of light and are kept away from the frost, they will start to

Spot the little treat for me!

## POTATO COMPOST

For every 15 litres of my own sieved soil I add:

- 10 litres rotted horse manure
- 15 litres compost from out-of-date growbags
- 100 g fish, blood and bone
- Handful of organic potato fertiliser

*Divide roughly between three large plastic trugs then pour backwards and forwards between them to mix it up.*

grow little shoots. It tricks them, see; they think it's spring. If temperatures drop, I sometimes cover them with a towel and I have a small heater in the Cave just to keep them ticking along.

- In late February I plant up my first tub in the greenhouse – usually Rocket potatoes, the one that started my Twitter journey. I plant up another tub in the second week of March.
- First, I put 10 cm of compost on the bottom of the tub, and place two seed potatoes opposite each other on the compost, making sure the strongest shoots on the potato are facing upwards. Then, I sprinkle mycorrhizal fungi powder over and around the potato tuber. Then, I fill the tub to about 10 cm below the rim and push in two more potatoes, making sure they're not planted directly on top of my first two potatoes. I cover these with more compost, to 5 cm below the rim. This leaves room to water without overflowing the top of the tub and washing compost away.
- The tub goes in the corner of the small greenhouse. They don't need water at this stage. The compost I've mixed should be damp enough for the potatoes, until they start showing above the surface, and potatoes themselves are full of moisture. Over watering at this point could cause the seed potatoes to rot.
- I start planting up the outside tubs around 12 March, at the same time as I am planting them in the ground. I have an area in the garden where I put a double row of tubs. It takes me up to a week to plant them up.

- The tubs have holes in the bottom, to allow the roots to grow into the ground, so I make sure to dig the ground over and sprinkle fertiliser over it. This will help to feed the plants later when the roots find their way through the holes and into the ground, and should increase the crop.
- Once the shoots start showing, it's time to keep a really close eye on the weather forecast. If there is a suspicion of frost, get those tender shoots covered. I have pieces of hessian sack that I put over when necessary – it's the same stuff that stonemasons use to cover up freshly concreted walls to stop the concrete getting frozen and ruined. If you don't have access to this, then horticultural fleece will do the same thing. I remember in my dad's day we would cover everything with pea sticks or any other bits we had been pruning. The theory is that the twigs get the frost and protect your plants. We've moved on to more modern methods nowadays, but I don't knock it, it worked for Dad!
- Once they're growing well and all fear of frost has passed, I put a row of canes either side of the tubs so I can keep the foliage a bit neat with twine, and stop it flopping everywhere.
- Then I just keep an eye on them and water as necessary. I think a good soak once a week is definitely more beneficial than a dribble every day because it encourages the roots to grow down to hunt for water and nutrients. If you water every day, the root system will never need to go deep to find them.

**TIP**

Garden centres sell out-of-date compost very cheaply! I've been known to pick up a bag for 50p. I say, 'a bag', but I actually bought fifty bags when they were going that cheap! I put it on my compost heap and 'reheat' it, killing any nasties and making it good to use again. Then I just sieve it again, when needed.

- I also like to keep everything neat around them because a messy garden is more likely to have pests and diseases than a tidy one.
- Most important of all: keep an eye out for blight. Potato blight is a fungal disease that is spread by the wind. It was the cause of the Irish Potato Famine in the mid-nineteenth century. It's a real problem that everyone has to take measures to guard against. There are more resistant potatoes around now, but it does still show up, every now and then, and can devastate your crop as the tubers rot in the ground. If I see any brown patches on the potato leaves or they are turning yellow, I'll immediately cut the leaves and stems off and burn them. If you can't burn them, put them in landfill, never throw them on the compost heap or in the garden recycling bin.
- When it's getting close to harvest time you can often get an idea of what you might have by running your fingers through the soil just around the edge of the tub to see if you can feel potatoes there. I often check the outside of the tubs too, as you might notice bulges where potatoes are actually pushing against the side. This gives me an idea if it's going to be a good crop!
- When the big day arrives, I am very excited. The first harvest is the Rocket potatoes in the greenhouse. Liz and I love to see what we've got and Liz usually takes pictures or does a short video of this. I pull the tub out of the greenhouse and cut the foliage off. We put a small tarpaulin on the path, then tip the tub upside down and tap the compost and potatoes out.

## When I harvest my potatoes

**First Earlies**
70–90 days from planting to harvest

Harvest 1–2 weeks from when they start to flower.

**Second Earlies**
90–110 days

Harvest about 3 weeks from when they start to flower.

**Maincrop**
At least 120 days

Harvest when the foliage has died down.

- This is the really exciting bit: the moment when you see if all the work you have put in is worth it. Carefully picking every potato out of the compost, I put them in a pre-weighed bucket and, when I've got every last one, I weigh them. If I'm lucky, it will be a new record for me. Then they are taken indoors and eaten with our lovely salads over the next few days.
- This goes on periodically over the rest of the summer and it's always exciting. It's like having Christmas several times over!

**Heaviest potato**
4.98 kg
UK, 2011

**I've always grown a variety of summer and winter squashes for eating.** After all, variety is the spice of life! But trying to grow a giant squash is a whole new ball game. I say giant, but I'm only a beginner compared to the big boys. My biggest pumpkin weighed 80 kg and my biggest marrow was over 20 kg. These are just tiddlers for them.

When I plant marrows, I always think of my dad. He'd wanted to grow a large marrow, but feeding his family was more important. When I was young, I thought his were enormous but they probably weren't more than 7 kg. So I'm doing this for you, Dad.

Pumpkins and marrows really need lots of space. I've contemplated digging up the lawn to make room, but I don't think Liz would be too happy with that!

I tend to grow one or two new things every year. Last year it was a tromboncino, an Italian squash that grows long with a bulb on the end, like a trombone. It needs strong support, so I have built a permanent frame for it up the side of our small shed. It runs along the boundary between our house and our neighbour's. It became a bit of a talking point for him and his friends! My best one was 178 cm long, which is pretty good, but I'll keep trying to do better. I'm growing a different variety this year called a Snake gourd. They're about as long but more wiggly.

## How I grow squashes to show

**Recommended pumpkin variety for competitions**

Atlantic Giant

- Marrows, pumpkins and tromboncino are all tender, which means if they get any frost at all they will die. As a result, I don't start to sow any of them in the greenhouse until late March or early April.
- I sow the seed individually in 7-cm pots and label them up. After watering, I put them in the heated propagator till they start to grow and then move them onto the heated bed. The propagator is slightly warmer than the bed, and has a lid, which keeps the humidity up and stops the compost drying out too quickly. I make sure the young plants are well watered but not soaked so much they drown. When the compost looks a little dry, I put them in a gravel tray and water from below until the compost looks damp.
- It's time to pot on into a 15-cm pot when they have grown a good root ball but are not pot bound. A plant becomes pot bound when the roots try to expand to find more nutrients but end up winding around themselves in circles; there are too many roots in the pot and not enough soil to hold and distribute water.
- To check the root ball, you need to see if there are roots coming out of the holes in the bottom of the pot. If there are, gently tip the plant out of the pot. Carefully tease loose any roots that are curling around and sprinkle them with mycorrhizal fungi powder before replanting.
- Before I move the plant I take the 15-cm pot and an empty 7-cm

pot and put enough compost in the bottom of the larger pot so that when I put the smaller pot in the middle, it sits a little way below the rim of the larger one. Then I fill both pots with compost and tap gently to settle it. After that, I remove the smaller pot, leaving a hole that is just the right size for the plant to go in. Then I put mycorrhizal at the bottom, before placing the plant in the hole and firming it in.

- The pumpkins and marrows go on the allotment once all risk of frost has passed, and the tromboncino is out in the garden. Most years, this is in the first week of June, but I always check the forecast because occasionally there can still be a frost.

- Before planting them out, I prepare the ground because they all grow so fast and are so big. They are very hungry plants that need lots of nutrients in the soil. Every third autumn, I grow green manure over the whole space where I grow the marrows and pumpkins. Green manures are plants that gardeners often grow and dig back into the ground to help keep the soil healthy. I plough mine back in with an attachment on my rotavator.

- In early spring, I dig a large hole and tip in six barrowfuls of rotted horse manure. After that, I put the soil back on top, which gives me a slight mound in the ground. This is where the marrows will be planted. Leaving plenty of space, I then dig another hole and put two barrowfuls of manure in. This is for the pumpkin. I follow the same process for the ground beneath

**Heaviest marrow**
93.7 kg
Netherlands, 2009

**Heaviest pumpkin**
1,190.49 kg
Germany, 2016

my tromboncino supports, preparing the ground with a couple of barrows of manure. By June, I'm ready to plant out. In the bed I put one marrow plant at each end of the large mound and one pumpkin in the middle of the smaller mound. I plant out two tromboncino, one at each end of the support.

- As the plants grow, I visit them at least once a day. Once I see the fruits are starting to grow, I cut all the other flowers off the plants. I want all their energy going into just one squash. As the marrow gets bigger, I put sand underneath it to keep it from touching the ground. This also helps to deter slugs as they don't like sand. With the tromboncino, I wouldn't choose a flower until they are at least 1.5 metres off the ground, because it is a very long squash.
- At the height of the season, I water them all every day. Once a week, I add a high-potash soluble fertiliser to the water.
- As the marrows grow, the leaves take over a large patch of ground. This is good – the more leaves, the bigger the end result! As they spread, I put some compost over the vines at several places to help stabilise them. If they are covered, they will grow more roots which will help feed the plant and ultimately the fruit.
- As the marrow fruit grows, I put an old fishing umbrella over it to keep the sun off. This may sound a bit silly, but it stops the skin going hard which would in turn stop the marrow growing any more. In other words – don't let your marrow get sunburned!
- When the squashes are beginning to get really big, I measure

**TIP**

Marrow leaves are very susceptible to powdery mildew but a good tip, which a lot of experts swear by, is to mix one part milk to ten parts water and spray it over the leaves.

If you don't have
an allotment,
rather than
growing a trailing
marrow type you
can try bush types
in your garden.
However, you
really need a big
space to grow any
pumpkins, because
even the smaller
varieties take up
a lot of space
— they're very
invasive plants!

them every day. By keeping track, I am able to tell if they are still growing. If the measurements stay the same for four days or more, I will cut the fruit off. Marrows can keep growing until the first frost of autumn, so if you harvest one in August, you'll still have a month or so to get another.

- Once I've got the biggest marrow of the season, I cut it off with 15 cm of stalk still attached, get it in a wheelbarrow and take it to be weighed. Fingers crossed it becomes the biggest one I've ever grown.

- With the tromboncino it's the length that matters. I didn't weigh my biggest last year, but it was 178 cm long.

- The only difference with the pumpkin is that it is so big, I need help to load it into my trailer and I have to take it to a weigh bridge. In the next village, there's a very old flour mill that has one, as well as big scales for weighing their flour. So I asked them for permission to take my pumpkin there. When my friend and I unhitched the trailer and pushed it in, the manager came out and saw us with this enormous pumpkin. I can't repeat his first words, but he seemed quite surprised at the size of the pumpkin! That one was about 68 kg, the size of a lorry tyre! Not a monster in the grand scheme of things, but big for me.

**Cucumbers are one of my favourites to grow big and can be sown indoors or outdoors.** Unlike the other cucurbits, I grow mine in the greenhouse for the most stable conditions, but you could grow them outdoors in a nice warm and sheltered spot – although it does depend on the variety. Make sure you check the guidance on your seed packets for the best results. The seed I use is an unnamed variety that I was gifted by a friend, but there are plenty of options to choose from in seed catalogues.

Those plants can really take off! I'd like to try growing two of the large cucumbers, but I can't because they'd take up the whole greenhouse and I've got so many other things growing in there.

I harvest giant cucumbers when they stop growing, which, if I've got my maths right, should be the day before the show. Everything has an optimum growing time so I have to work back from the show date to know when to start sowing.

There was just one national CANNA Grow Show in 2020 because most of them were cancelled due to Covid and I got a third with a cucumber that was 71 cm long and 28 cm around. You can see it at the bottom of the photo, nearly falling off my knee. Liz and I made a lovely cucumber-and-sweet-pepper chutney with it.

## How I grow cucumbers to show

- I grow my cucumbers in the borders of my greenhouse where I have a sturdy framework for them to climb up. I make sure to prepare borders with plenty of rotted horse manure incorporated.

- I sow the seeds in the first week of May. I plant two seeds at the far corners of each border. This is just an insurance policy in case a seed doesn't grow. Once they are growing, I will remove one seedling from each border leaving one plant either side.

- After a slow start, the cucumber plants will suddenly speed up and throw flowers everywhere. Now I'm looking to choose good looking fruit that has been pollinated. You can tell when pollination has happened because the flower drops off, leaving a healthy little cucumber behind. If no pollination has occurred, the fruit falls off with the flower. Once I've made my choice, I pinch out the rest.

- One thing to keep an eye out for is hidden flowers. Cucumbers are masters of disguise and, unless you are very alert and diligent, you will find another cucumber hidden amongst the foliage taking nourishment from your precious baby!

- I keep close watch on that cucumber; if it stops growing, I will have time to remove it and start again. I feed the cucumbers with Tomorite at the same time as I feed my tomatoes. When they start to grow big I measure them every day, so I know when they've stopped and are ready to harvest.

You have to hold the cucumber up with a frame and a sling, otherwise it will rip away from the mother plant.

**Runner beans are so easy to grow that even a beginner can get a good crop.** I believe that anyone with a garden should grow a runner bean.

For me, just the words 'runner beans' conjures up summer. They're delicious to eat and both Liz and I love them. Last year Liz made a lovely runner-bean chutney, with spice paste in it. It was beautiful with a curry. The crop can go on for a few months, so for the ones we eat, I pick them when they are young and tender.

If you're looking to grow long or large runner beans, you will need the right seed. I've got a couple of new varieties on trial this year that I'm growing alongside my favourite, Guinness World Record. My longest bean is 72 cm. The UK record is 86.5 cm. It might sound like I'm very close to the record, but the difference is massive.

Seeds can be expensive, so for runner beans I try to save my own seeds For these I leave a few large pods on the plant to grow on once I've finished harvesting them. In the first week of November, I collect the dry brown pods and bring them inside. They take a few more weeks to completely dry out, but when they are ready I can hear the beans rattling around in the brittle pods. Then I take them out of the pods and store them in a cool dark place in my Cave. Look after the pennies they say!

# How I grow runner beans to show

**Recommended variety for competitions**

Guinness World Record

- Digging my trench for the summer's runner beans is one of my first jobs of the year. The trench is around 60 cm wide, 30 cm deep and about 5 metres long. I make sure to leave space on both sides so that I can walk along for watering and harvesting.
- Once I've done the digging, I throw in any vegetable scraps from cooking. Those scraps go in the bottom of the trench and rot down. You can also add eggshells and used coffee grounds.
- I start my runner beans off in the greenhouse at the end of April. I literally throw a handful of seeds into a large pot of compost, cover them over, and let them get on with it. If you want to, you can sow the seed straight in the garden, but you need to sow a little later, though – on 12 May. Well, that's the date I would sow outside, anyway.
- Next I put up my bean supports. I put hazel poles or canes in every 30 cm along both sides of the trench. Then I join them at the top, almost as you'd imagine a wigwam or a tent. It's not essential to do this in advance, but I like to make sure it's done so I'm never rushing.
- I plant the seedlings out at the end of May. I plant one runner bean by each stick. By the time you plant your beans out, they will only be about 10 cm tall. But runner beans are incredibly fast growers, and it won't take long for them to get going.

- I tie the plant carefully with a piece of twine to the stick, then keep an eye on it for a couple of weeks, encouraging it along. Once it's wrapped itself around the stick once or twice, you're in business, and the rest should take care of itself.
- One thing to remember is that runner beans grow anti-clockwise. Most things follow the sun, including my dog and my cat! But runner beans go the other way. When the sun's out in the summer, I will know precisely where they'll be at any given time, because they'll be trying to get as much sun as they can.
- So, when you're tying your bean, just make sure you gently wind it anti-clockwise up the stick. If you wind it clockwise, the bean won't grow, it'll just unwind itself and fall back down. When that first happened to me, I had no idea what was going on!
- To be on the safe side, I put fleece around the bottom of the sticks to form a barrier all the way around in case there is a frost. The fleece stays on till the second week of June.
- By now they should be growing well. Beans are very hungry plants, but as long as I have made my trench properly, there should be enough nourishment there to see them through the season. However, in order to make the best of the food, they need water. I make sure they don't suffer in a dry spell and keep them well watered – giving them a good soaking once or twice a week.
- Also, if needed, I give them a bit of help to make sure they're actually climbing up their poles and not along the ground. Runner

**TIP**
If you haven't got room for a row of runner beans, why not try a wigwam? Put six or eight sticks in a circle, and tie them together at the top. Then plant one bean at the bottom of each stick.

beans are what I call a 'flippity-floppity' plant; they'll fall all over the place if they don't have the extra support!

- When beans reach the top of their sticks, I need to pinch them out. That means I pick the growing points out, to stop the plant getting any taller. It's also a good idea to spray the beans with water as this aids pollination.
- As the plants grow, they throw a lot of flowers and beans. We pick and eat them every day until I notice a good one about 1.5 metres off the ground. I'll let that one grow, and hopefully another one will come along, too. Everything else is picked off but once these two big beans start really growing, the plants tend to put all their energy into them and will stop flowering and fruiting. After all, this is what they are there for – to produce seeds.
- It's important to remember that the steps I have described are for growing a really long bean. A show bean will ideally be long, but also needs to be young and tender. You will need between five and ten beans, depending on the show, the class, and its entry requirements.
- If you are entering a show, it's important to pick beans that are as identical as possible. Something else to note is that when you come back in after the judging, you may find that one of your beans has been snapped in half. This isn't vegetable vandalism; this is part of the judging process, to check they are tender and can be eaten!

WORLD RECORD

**Longest runner bean**

130 cm
USA, 1997

**Although I love the challenge of growing big veg, I love growing food to eat even more.** Liz and I love our veg and our salads, and we're eating homegrown produce every day. Even when there's nothing to harvest we'll be digging into our stores of chutneys and pickles. At other times of year we're harvesting so much, we just can't eat it all ourselves, but we love to share what we grow with family and friends, and the local community. We have two nursing homes in the village so sometimes I'll pop over to them and drop some veg off. I also try to plant little and often, and this means we can be eating some things nearly all year round. If we're lucky we can have lettuce from March till October, and I always make sure to plant some Charlotte potatoes to be ready on Christmas day.

Some times of year are busier than others, but I've never been one for sitting still. If there aren't seeds to be sown or vegetables to harvest, I'll be thinking about what I'm going to be planting next year and repairing the fences and sheds.

I've tried to give you an idea of what I'm doing in the garden – and eating – each month. I've included lists of what I'm sowing, planting and harvesting each month so you can pick the things you like to eat and have a go with these. I hope it encourages you to and try some veg growing of your own.

Cheers!

**January is a busy time in the garden; there's something to be done every day.** But I enjoy getting outside and being active after the Christmas period.

It's always going to be chilly out in the garden at this time of year but I don't let the short days get me down; I love hearing the robins sing in the garden on a frosty morning. Thankfully, in this day and age, warm clothing is so easy compared with a few years ago. I like my hoodies because I wear caps all the time: flat caps usually but also baseball caps. A cap is fine, but then if you stop for a break and your ears are a bit chilly you've got the hood to just pull up. When it's really cold I wear a beanie, but I do find that if I'm doing physical work like digging the soil, a beanie is a bit too warm.

I get mentioned for wearing Crocs all the time. I've got about five pairs in different colours; I find them so comfortable, even for gardening. I will only put wellies on if it's really necessary, even in the winter. I've got an expensive pair of wellies that I think are too good for gardening. I keep them for holidays. I don't want to wear them out!

Digging your plot in January is good preparation for planting in the spring. The only piece of ground on my allotment that I don't dig over is where I still have produce growing. At this time of year, most of my veg patch is

**TIP**
You shouldn't dig
if the ground is
frozen or if it
is too wet and
muddy. If you're
walking all over
your soil when it's
sodden, you can
press it down too
hard and the soil
becomes rock-like
when it dries out.

usually covered with black polythene. In the autumn I will have dug in green manure and covered it with black builder's polythene to keep it as warm as possible. Covering with the polythene helps to stop the rain and snow from leeching out all the nutrients the manure will create. But if you haven't dug your plot already, don't worry. January is a good time to get started on digging, so the wet and frost can get in and help to break the soil down. Start early!

There are a few odd jobs I like to stay on top of this month. If you haven't managed to clean all your pots and seed trays in the autumn, then it's a good idea to get it done now before the growing season starts.

I also like to check for holes in the nets over my winter brassica. These include cabbages, cauliflower, kale, white and purple sprouting broccoli and Brussels sprouts. I do this for two reasons. The first is because I don't want a bird to get tangled in them – I also have my nets on frames so that there is less chance of that happening. The second is because I don't want to share my greens!

I love birds and put a lot of food out for them. But I like to keep the green stuff for myself.

Birds are definitely friends of the garden. Often, when I'm digging with a fork, I will have a robin or a blackbird a couple of feet from me, waiting for the worms and creepy crawlies. But, if you're not careful, the birds will eat some of what you are growing.

It is also worth regularly checking for slugs in cabbages and cauliflowers. They often hide in the outer leaves and, in the milder weather, will have a feast.

A big job for me this month is digging my trench for the runner beans (see page 102). You don't plant until May, but it's very important to prepare the ground in January.

At this time of year, it's also important to check any potatoes that are in store. However carefully they're stored, there's always a chance that one has started to rot. If it's not removed you could lose a lot of your spuds, which could mean not having enough potatoes to last until spring. I aim to put away enough potatoes in the autumn to keep me and Liz going till late April or early May, by which time the early potatoes planted in my greenhouse will be ready to harvest.

At the same time, I am 'chitting' my seed potatoes ready to plant from late February onwards. I put my seed potatoes on to egg trays in my Cave so they will think it is spring and start to grow shoots from the 'eyes' (see page 80).

If you have a heated greenhouse, January is the time to start sowing your seeds. Some plants, like aubergines and sweet peppers, need a very long growing season in order to reach their full potential.

**Sowing in January**

Aubergine

Sweet pepper

Onion

Celeriac

Chilli

Spring onion

Tomato

If you don't have a greenhouse, you can start off sweet peppers, tomatoes and chillies on a windowsill indoors. You can actually buy a propagator to fit on a windowsill. Just don't forget to take the seedlings out as soon as they start growing, or they may become leggy — by that I mean very long and weak instead of strong and sturdy. It's more important to have strong root growth than strong foliage at this stage. Even if you have a heated greenhouse like I have, a heated propagator is definitely a welcome addition.

Be careful when choosing your chilli. A couple of years ago, I made the mistake of growing Trinidad Moruga chillies. The Scoville Heat Scale runs from 0 to over 3 million heat units. I knew they measured a whopping 2 million Scoville Heat Units, and the packet recommended handling them with gloves, so I knew they'd be hot. I like a bit of spice in my food, but I hadn't expected quite how fiery these little red scorpion chillies would be. I tried one raw and as soon as it touched my tongue, I knew I'd made a big mistake. I was coughing and spluttering! I found out later that one small Moruga chilli, about the size of a golf ball, is enough to flavour a thousand bars of chilli chocolate. Nowadays, I grow stuff that's a little more palatable!

There's lots to enjoy from the garden at this time of year: parsnips, late-grown carrots and other root veg such as

## My favourite chillies

### Cayenne Chocolate
– A unique variety that acquires a deep chocolate brown colour before turning red. They look like little chocolate chillies for a while. They go well in my chilli jam, as well as chutneys.

### Trinidad Perfume
– A mild, sweet, and slightly citrussy chilli. They are a bright yellow and grow into a delicate lantern shape.

salsify. Salsify grows about a foot into the ground, it's carrot shaped and covered with lots of hairs. You might recognise its delicate purple flower. It's a lovely vegetable, similar to a parsnip, which I often use in my slow cooks.

There are also Brussels sprouts and other brassica that can withstand a frost. One of my personal favourites is cavolo nero, an Italian kale. It's almost blue in colour, but they call it black kale. When I mentioned cavolo nero on Twitter, somebody reckoned you should take the stalk out and sprinkle rock salt over the top, along with some olive oil, and put it in a hot oven. That way you make it crispy. I rather like the sound of that. Once I blitzed it up and made a smoothie. I felt very healthy drinking that, but it was a little bit strong flavour-wise. Cooking removes some of the bitterness so I tend to keep things simple and steam or boil it.

First of all, I take the leaf, and with my finger and thumb I rip the leaf from the stem. I'm left with the leaf in one hand and the stem in the other. I throw the stem in the compost. I have seven purpose-built compost bins. Nothing is wasted! Then I slice the leaf into thin strips and steam it. It's beautiful, and very healthy. We generally have it with things like pie and mash, or with a casserole. I think it's one of those vegetables you can have with anything.

**February is a busy month on the whole, with lots of preparation work for the coming spring.**

It can be a rotten month for weather. When I was younger I worked on the River Thames as a bargeman. When the river was in full flood at this time of year, I would have to rescue people who were stranded on their boats. Thankfully I've not had a bad experience of the February weather in my Cotswolds garden yet, but at this time of year I always check everything is secure, just in case. If I know there are high winds on the way, I also check that my fences and trellis are in place. There'd be nothing worse than waking up to find half a fence had blown down and broken.

It's also a good time to do a quick audit of my cold frame and cloches to see if they are in good order and that there are no breakages, and to do any running repairs. I also make sure the metal legs of the cloches are firmly in the ground with no space for the frost to creep under.

In February, we start to notice more birds coming into the garden as they gear up for the mating season. We have a good selection of birds that come and go from the table that I built. Every day I'll see blue tits, robins, starlings, and goldfinches, too. They're very pretty with little red faces and a yellow streak in their wing. Their song is beautiful, too. We see the occasional long-tailed tit. They don't come every

day, but two or three times a week. I like to watch them swoop in and perch on the table. Watching them while we have our lunch is better than any TV show.

As February can be very wet and windy, I try to keep off the plot until the weather is dry enough for me to walk without the ground sticking to my Crocs. In January and February, I only walk on muddy ground to harvest things.

When the ground is dry, I like to spread a layer of manure on most of the beds. Manure encourages the roots to fork so I don't lay it where I will be planting root vegetables. I use my rotavator to make sure the manure gets into the ground really well, but if you haven't got a rotavator, you can dig it in.

I have several rotavators. I can't resist them. It's like a hobby of mine. I think in every man there's a little bit of an engineer: a fascination with machines and gadgets, things that you can start up and watch whirling around and coughing out smoke. I think most men like that stuff. A rotavator is the closest thing you can get to a tractor. I have tried to persuade Liz to let me have a small tractor, but she makes the very valid point that it wouldn't fit through the garden gate. That's a shame, really. I'd love a tractor.

There are a few other bits of housekeeping that I make sure to do in February. I find that it's a good time to check

**Sowing in February**

Aubergine

Lettuce

Small cucumber

Sweet pepper

Tomato

**Planting in February**

Potato

## My favourite lettuces

**Analena** – A tasty lettuce with a huge head.

**Lobjoits Cos** – An old favourite, supposed to be the best Cos.

**Oakleaf Navara** – This one is really deep red.

**Red Salad Bowl** – A great one to grow as a cut-and-come-again variety.

**Webbs Wonderful** – A good old-fashioned lettuce. If you like Iceberg lettuce, this might be the one for you.

how much compost, fertiliser and liquid feeds I have, and order any extra I may need before the growing season really takes off. Believe me, it's really annoying to go to sow seeds or prick out seedlings only to find that you've run out of the compost you need halfway through. I've been there and got the badge.

If you haven't already done it, now is a good time to clean and oil your tools ready for the spring. Don't forget to sharpen any tools that need the extra attention. I also find it a good time to tidy up my Cave before things get really hectic in the garden.

Around this time of year, I prepare the 200-litre barrels that I grow my long carrots (see pages 38–47), parsnips (see pages 48–53) and beetroot (see pages 54–57) in.

It's also time to sow my cucumbers. As well as the big cucumbers I'm sowing for the veg shows in the summer (see pages 96–99), I sow my Mini Munch cucumbers. The Mini Munch grows to a lovely, dinky size. You can eat them in one go. They're so easy to grow – you could have one in a pot on a veranda or in a conservatory, and when it's ready, you just crop it. The more you pick, the more will grow.

I sow two Mini Munch seeds in a 7.5-cm square pot, pushing them into the surface of the compost and covering with vermiculite. I sow two because you never know if they are both going to germinate, so it's a bit like having an heir and a spare! Then I put the pot into the propagator in the large greenhouse until the seedlings show above the surface. At this point, I put them on the heated bed until they have their first 'true' leaves. This refers to the second set of leaves. The first set are called seed leaves. Once I see the true leaves I gently tip the pairs of seedlings out of their pots and plant each separately in its own 7.5-cm pot.

Now is the time I sow my first lot of lettuce seed. I always make sure to sow a variety as we like to mix and match. I get them started in the greenhouse by putting just a few seeds in a small container. About a week after they are sown they are big enough to prick out into modules, where I leave them until they are 5 cm tall.

In the last week of February, I plant a tub of first early potatoes (see pages 80–81). I usually choose Rocket and plant them in a 30-litre container, which I can buy at a reasonable price online. Being a pensioner, I do have to watch the pennies.

In February we continue to enjoy lots from the garden. We'll still have plenty of cavolo nero, parsnips salsify and a few carrots. We'll also be anticipating the purple and white sprouting broccoli . . . two of my favourites!

## Harvesting in February

Carrot

Leek

Parsnip

Salsify

Brussels sprouts

Cavolo nero

**March is a special month for us, as Liz's birthday is on the 24th.** I'll always make or repair something for her, it's too easy just to go and buy a present.

It's a very unpredictable month, weather-wise. The old saying is 'in like a lion and out like a lamb' – meaning that the weather is rotten at the start of the month and quite pleasant by the end. I always hope March comes in with a gale. If it comes in like a lamb I know we're in trouble, because it'll go out like a lion. I'd rather have the rubbish weather at the beginning of the month and better at the end.

I make sure to watch out for pests as it gets warmer. If it's a damp mild night, I sometimes go on a slug and snail hunt with a torch and a container to put the prisoners in. It's me or them! What other people do with them after they've caught them is their decision. Some people like to take them for a long walk and release them far away. I tend to take a more permanent solution to the problem . . .

In March, there is so much to fit in that sometimes it feels like there aren't enough hours in the day. If the weather is good, now is the time to cultivate the soil. I will dig if necessary, then run over the soil with a rotavator. I also hoe, weed, and rake my proposed seedbeds, then cover them with cloches or polythene. It's very important to try and warm the soil as it can still be a bit chilly.

As part of my preparation of the soil, I check the pH levels, because certain vegetables, like brassicas, prefer more alkaline soil and others, like onions, prefer an acidic soil. If needed, I will add lime to make the soil more alkaline, or sulphur chips to make it more acidic. It might sound very scientific, but it's not, I promise! It's very easy, once you get the hang of it.

It's also a good idea to spread some fertiliser over your cultivated ground, except where you are going to sow your seeds. Seeds don't want too rich a soil to begin with. I tend to use fish, blood and bone, but any general fertiliser will do the job.

If you're new to gardening, you might be surprised to hear about fish being used as a fertiliser. In fact it's very common to use sterilised fish, just make sure you don't use ordinary fish. Trust me. Back in 1976, I worked at Farmoor Reservoir. My job was to feed the fish. One day I found some out-of-date trout pellets, made of fishmeal. I thought, 'I'm going to dig those into my plot. I'll have monster potatoes.' The potatoes I grew were fabulous and I was so proud. I took them inside and my then wife said we'd have them that afternoon. I went back out to the garden with a big smile on my face. But a little while later, she came out and said, 'There's something wrong with those potatoes, the kitchen stinks of fish.' I'm afraid she was right, the potatoes had taken in the taste of the pellets. I was devastated. In the end, I had to give two rows of beautiful spuds to a local farmer to feed his pigs.

In the large greenhouse it's time for more seed sowing. About the second week of March, I sow various cauliflowers, Brussels, leeks and cabbage. I'm also sowing celery – I grow a variety called Mammoth Pink.

As the old saying goes, 'don't put all your eggs in one basket'. In this case, that means don't sow all your seeds at once. There's always a possibility that your seeds fail to germinate. It can happen for any number of reasons and sometimes it's hard to put your finger on why. One year every seed grows, the next only one or two. If you stagger the sowing of them, then there are more seeds to try later. So to be safe, I only sow half a packet of seed at a time.

It's time to think about my veg for the summer shows now too. I will have moved my big onions to 11-cm pots at the end of February and they will continue to grow throughout March (see pages 68–77). The middle of the month is when I sow carrots, parsnips and beetroot in the barrels I've prepared. I'll also be continuing to sow tomato seed and last month's sowings will be ready to prick out (see pages 78–87).

## Planting out in March

Beetroot

Carrot

Lettuce

Onion

Parsnip

Potato

Tomato

By the end of March – weather permitting – I plant all my first early potatoes in tubs outside (see pages 78–87) and in the ground. To plant in the ground, I use a 'dibber'. I make a hole with the dibber about 15 cm deep then drop the seed potato in, making sure the strongest shoots are facing up and it is in contact with the soil. It's important there is no airlock underneath. Then I fill in the hole. I leave about 30 cm between each potato and 60 cm between each row.

Once all my potatoes are planted, I use organic potato fertiliser. You can buy a bucket of 10 kg for about £20. I sprinkle it by hand over the ground. Then I fetch the hoe and run it up each side of the row with to 'baulk up'. This basically means drawing up the earth, so each row will be covered with a small mountain of soil.

March is the time to plant Jerusalem artichokes: a vegetable that is part of the sunflower family and that you rarely see in the shops. They tend not to be too popular, as they can become quite invasive if you don't keep them under control. The underground tubers are eaten from October to March, but are at their best between November and February. Liz and I love them.

At the beginning of the month the first of my lettuces should by now have reached 5 cm and are ready to plant in their final position in the salad bed. I put a cloche over any

**TIP**
A dibber is simply a piece of wood used to make a hole – ideally with a handle at one end and a point the other. An old wooden garden fork handle makes a great dibber.

My homemade scoop made from a cider-vinegar container.

lettuce or salad leaves that I plant out in March or April, just to make sure I keep them warm.

We are lucky enough to have inherited an asparagus bed on the allotment. Shop-bought asparagus just doesn't compare with the flavour of fresh-cut, homegrown asparagus, lightly steamed and covered with butter and a sprinkle of Himalayan pink salt. This is the time I weed it. I don't dig too deep as the roots are close to the surface and I don't want to disturb the dormant plants. I then top dress the bed with a growmore or fish, blood and bone fertiliser before the plants really start growing.

If you don't have a bed already, then March is also the time to make one – though it'll be a couple of years before you will be able to harvest it. You can grow asparagus from seed, but it's better to buy dormant year-old plants known as 'crowns'. Dig a trench about 20 cm deep with a little ridge along the middle for the crowns to sit on top of. Put down a layer of manure covered by a layer of compost. Do this properly because it will be there for twenty years! Then plant your crowns 35–40 cm apart, fill in the trench to cover the plants, and give them a good water to settle in.

We're harvesting white and purple sprouting broccoli this month. We tend to have it steamed, and it's absolutely delicious.

## Harvesting in March

Leeks

Lettuces

Sprouting broccoli

**In April, the garden comes to life as lots of Liz's flowers start to bloom.** The longer days and warmer temperatures do have a great effect on the garden. Everything starts to grow faster.

At this time of year, everything is growing well in the greenhouse too and it's getting pretty full. We have to spend a lot of time juggling our plants around. Liz and I are usually fighting over the space. It's between her flowers and my veg plants . . . I usually win, you can't eat flowers!

We have double-layered staging, so to maintain even growth it's a case of swapping plants from the top to the bottom every three days.

I do have to watch out that the greenhouse doesn't get too hot. Even though it's still chilly at night, you can easily be caught out by the speed it heats up on a sunny morning. It may not even feel that hot outside, but under the glass it can soon climb to 30°C. If there are any seedlings in the propagator with the lid on, they will fry, so I make sure I'm checking the temperature all the time.

We have to be very vigilant this month as it's easy to become overwhelmed with slugs and snails. As well as our evening patrols, we encourage the birds into the garden. Blackbirds will eat slugs, and we even have an occasional thrush visit us for the snails.

In the greenhouse we look out for ants. You may think that ants aren't a problem, but ants are farmers and what do they farm? Aphids! They guide the aphids to nice juicy shoots to feed, they produce a sweet sticky substance that the ants can't get enough of. And what's in a greenhouse at this time of year? Lots and lots of lovely juicy shoots.

This is a busy month for sowing if you haven't got young plants in the greenhouse, as the ground is drying up and it's now possible to sow seeds outside.

In the first week of April, I start with cabbages, cavolo nero and cauliflower in seed trays. I like to start all my brassica off this way, but you could wait till May and sow them directly outdoors. I'm also sowing white and purple sprouting broccoli. They won't crop till next March, but they need a long growing season. Sweetcorn is also sown in April so that it's ready to plant out in early June.

Also in early April, I sow asparagus peas. They are delicious (like a cross between asparagus and peas!) and as a bonus they have very pretty dark pink flowers. I would say that they're something that you've got to grow yourself as you probably won't find them in shops. We eat them steamed when they're about an inch long.

Then in the second week of April, I am busy sowing radishes, spring onions, kohl rabi, kale, spinach, salsify,

## My favourite cabbages

**Brunswick** – A winter cabbage that's great for making sauerkraut.

**Delight Ball** – A nice early cabbage.

**Dutchman** – A late-summer pointed variety.

**Golden Acre** – This one's crunchy and good for coleslaw.

Brussels sprouts and chard. You could sow all of these in seed trays, but we usually sow all of them outside. I cover the beds with horticultural fleece at first to stop the birds from scratching up the seeds. Once they have germinated, I can take it off. I will also sow carrots, parsnips and beetroot, including the ones I am doing in my tubs for the veg shows.

Back in the greenhouse, I sow runner beans and French beans. I sow them one to a module. There are twenty-four modules to a standard-size seed tray. They should be ready to plant out in the first week of June, when hopefully we have all the frost behind us.

Mid-April is also time to sow courgettes, marrows, pumpkins and other squashes. For the ones that are to eat I fill up 7.5-cm pots with compost, and sow one seed per pot. After watering, I put them in the heated propagator until they start to grow. They will be ready to put on the greenhouse staging by the middle of May.

In mid-April, I sow a couple more cucumbers and start them off in the propagator as well. It's a heated black tray, around 70 cm long by 40 cm wide with a clear plastic cover, which we keep in the greenhouse. It's important to check seeds growing in a propagator every day, because as soon as the seed pops up, you have to take them out, otherwise they get leggy.

**TIP**

Don't wait until weeds are knee high. Far better to get a hoe and go over the ground at least once a week to stop them before they start. This also allows air into the soil and makes the surface 'friable', which means easily crumbled.

So, as soon as the seedlings show, I move them to my heated bed. A heated bed doesn't have to be anything flashy. I built a box about 15 cm deep from 2-cm marine plywood around an aluminium base. Then I lined it with polythene and half filled it with builder's sand. On to this I laid a long soil-warming cable snake-wise with about 5–10 cm between each loop and then put another 5 cm of sand over the top. The sand needs to be kept damp at all times because this retains the heat better. After a year you might have to top up the sand as it settles down.

It's important that your cucumber seedlings don't get cold. A couple of years ago we bought two lovely, strong little plants and between taking them from the garden centre to the car, and from the car to the greenhouse, they must have caught a chill. Sadly, within days they died. So I'd recommend taking something to cover your plants with if you're buying in the colder months. It's not worth running the risk!

By April, lettuce and other salad leaves can be sown outside. I sow rocket, red-veined spinach, sorrel and lamb's lettuce straight into the salad bed.

The area for the seedbed should be roughly levelled, then trodden down and raked until the ground is level. Make sure all stones are removed and the soil is pretty smooth. This gives the seeds the best chance to grow.

When I'm sowing seed outside, I draw a 'drill' – a shallow indented line in the soil – across the seedbed. If it's dry, I water the drill or, if it's very wet, I put some sand in the bottom first. You can sow things in the seedbed and transplant them but these days I leave the plants in the rows I sowed them in and thin them out as necessary.

For the lettuces, I start off with one plant every 7.5 cm, then a month later I'll take out every other one. You'll be wondering why I don't do that in the first place. Well, if some of the little plants die off for any reason, there would be large gaps in the rows. This way, you ensure you have enough strong plants.

The salad bed was the first raised bed we built. It's around 90 cm wide and stretches from just outside the back door to Liz's shed – about 2.5 metres. On that bed we grow lettuce and salad leaves mainly. And lots of them. We have both to cut for most of the year – even on Christmas Day!

As soon as the first early potatoes show above the ground, they need baulking up to cover up any leaves. If the leaves are left exposed, you run the risk of the frost killing their foliage. Baulking is something that needs to be done regularly until there are no more frosts. Remember: the bigger the baulk, the more potatoes! The piled-up soil also stops light turning the potatoes green and inedible.

## Sowing in April

Marrow

Pumpkin

Radish

Rocket

Runner beans

Salsify

Spinach

Spring onion

Sprouting broccoli

Summer squash

Sweetcorn

Tomato

## Planting out in April

Beetroot

Carrot

Celery

Lettuce

Onion

Parsnip

Potato

Shallot

Tomato

In April I'll also make sure the potato bed for second earlies and maincrop is rotavated and made ready to plant. I'm lucky that I have a fairly large rotavator with different implements, including a baulking tool. This is shaped a bit like an arrowhead. I move the rotavator backwards and forwards across the area I want to plant. This forms ridges and furrows in the ground. I then plant the seed potatoes in the furrows. After planting them all, I go back over the ground 'splitting' the ridge. By the end, I have ridges over the potatoes and furrows in between. Job done!

Early in April I prepare my onion bed for my seed-sown brown and red onions and shallots, and move the plants into the cold frame to harden off. It's too much of a shock to take them straight from the greenhouse to the garden. By the last week of the month, I can start to plant them out.

By mid-April the tomatoes I sowed first should be sturdy little plants so I can plant one or two in the border of my greenhouse.

April is the time that I enjoy the last of the leeks. I love them in cheese sauce. I do sometimes show leeks but they're not a popular class down south, so I usually just use them in a mixed display. If I'm very lucky, there's also a bit of sprouting broccoli left to harvest. Finally, near the end of the April, I can start to harvest my asparagus – heaven!

## Harvesting in April

Asparagus

Leeks

Lettuce

Sprouting broccoli

**May can be a lovely month. Summer is coming and the sun seems to shine more than ever.** This is the month we normally start having family barbecues.

The days are getting longer, too. That is good news, as there is so much to be done in the greenhouse, garden and allotment. I like to make sure I'm very organised with everything in the garden and in my shed as we come into the busiest months of the year for planting the veg we love to eat, and for the veg shows.

It's important to keep an eye on pests, as they can gobble through half your stuff before you get a chance to! Slugs and snails are abundant in May and need to be dealt with. Pigeons and hares are also pests on the allotment. Last year I think the hares had more broad beans than we did! We have the odd deer wander in for a snack but, by and large, we get on with the wildlife pretty well.

Hopefully by now the broad beans that I planted in October will be growing well and even flowering. I make sure I keep them well hoed and weed free. Even more than the hares, the main threat to broad beans is from black fly, which can descend overnight and ruin your plants. In order to prevent this, early in the month I remove all the tops from the bean plants, because these juicy little leaves are the main attraction. It's often suggested that the tops

are delicious cooked. We tried them last year and weren't impressed . . . each to their own!

My tomato and cucumber plants are all in my small greenhouse by now, and the heating in there is set to come on if the temperature drops below 10°C as this would affect their growth.

Unless there is a sharp frost forecast, I will turn off the heat in the large greenhouse. The heated bed will have done its job through the winter and can be turned off as well.

When the weather is nice, the door and windows in the large greenhouse are open all day. By the end of May, they will be open day and night. Hopefully it'll be getting quite empty by now, so this is always a good time to give any pots a clean before they're needed again.

Virtually any veg you want to grow can now be sown outside in a well-prepared seedbed. I'll be sowing root veg such as swedes, turnips and beetroot at this time of year.

Mangetout and peas can also be sown this month. I grow a mangetout called Oregon Sugar Snap. They're perfect for when you're walking around the garden. I just pick a dozen and eat them straight away. I use a Dutch hoe and draw it through the soil to make a shallow trench then scatter peas sparingly along the bottom before raking the soil back over. Most peas need support as they grow. I use pea sticks made from coppiced hazel branches. Or if I can't get enough of these, I buy 'pea nets'. I poke bamboo canes into the ground at each end of the trench and every 60 cm along its length. Then I put something on the top of each cane to stop the net slipping down – a yoghurt pot upside down works nicely. Then I drape the net over and pin it down so it doesn't blow away. The peas grow up it very happily.

We usually have a salad every other day, so we need a lot! With lettuce, I carry on sowing some under cover every fortnight and then planting them out once they're big enough. When the seedlings have grown to around 20 cm, they need thinning out to at least 40 cm apart to give the ones left room to grow. Some of the other leaves, like rocket and lamb's lettuce, prefer cool weather, so I stop sowing until the end of August. If it's too hot they just won't grow.

Spring onions are another favourite. I start my first batch in modules in the greenhouse as early as January and carry on sowing until the end of March when I plant them out. After that I sow a short row every couple of weeks until July. That means there's always some to eat. I am sowing radishes about once a week too now. Radishes are one of the fastest growing salad veg – they are ready to eat in four weeks, and I can have them from February to August.

## Sowing in May

Beetroot

Cucumber

Lettuce

Peas

Radish

Spring onion

Winter squash

Swede

Turnip

If you want to have more length of white on your leek make a baulk — the same way as for potatoes (see page 127) — then dib the holes in the furrow created. As the leeks grow, gradually fill in the furrow and even mound up the rows.

There's a lot of planting out going on this month including the runner beans that I am growing to show (see pages 100–105), various cucurbits – marrows, courgettes, pumpkins, summer and winter squashes and gourds such as tromboncino and Snake gourd – and all the brassicas. I just make sure I've got some horticultural fleece handy to protect the plants if there's a frost forecast. I leave the asparagus peas till the very last week of May – and even then, if the weather forecast is for freezing nights, I will hold fire until June.

At the beginning of May, when the seedlings have several leaves, I plant the celery I sowed in early April into 7.5-cm pots. To help provide the best conditions, I make a special raised bed beside my greenhouse. On that bed, I incorporate lots of well-rotted manure and homemade garden compost and add some growmore fertiliser to the surface. By the end of the month I can plant the celery in the bed. In the wild, celery grows in boggy areas, so it must never be allowed to dry out. I plant them about 30 cm apart and put a strip of damp-course material (plastic sheeting that is used in house building) around each plant. My dad used to cover his with corrugated cardboard, which also works well but sometimes can fall apart. Covering it like this helps to 'blanch' the celery. Blanching means covering the stems in order to keep the light out. It reduces bitterness and keeps the plant white.

May is when I like to transplant leeks outside. I find the best method is 'topping and tailing'. First, I take my dibber and make a hole about 30 cm deep. Next, I take a leek seedling, which should be strong by now. Then I trim the roots and cut off the tips of the leaves. Once I've done that, it's ready to drop in the hole. I leave about 15 cm between plants and 30 cm between rows. Then I just fill the holes with water. I don't push soil into the holes, I let nature take care of that.

At the end of the month, I plant out the sweetcorn I've been growing under glass, which by now should be about 25 cm tall. Because sweetcorn is wind pollinated, it needs sowing in a square block, not a row, so I leave about 35 cm between plants each way Then whichever way the wind blows, the pollen will do its job.

The potatoes need baulking up again now, to protect the leaves from a late frost and to make sure any baby potatoes don't see the light of day.

By May, my asparagus is coming on in leaps and bounds. I think they look like little fingers growing out of the ground. It's wise not to harvest for the first two years the plant is in the ground to give the plant a chance to become strong. But

## Planting out in May

Asparagus peas

Cabbage

Carrot

Cauliflower

Celery

French beans

Kale

Leek

Parsnip

Runner beans

Spinach

Sprouting broccoli

if you're a little impatient (and you love asparagus like me!) then you can start harvesting it after two years.

As soon as my asparagus spears are 15–20 cm tall, I cut them with a sharp knife just below the surface of the soil. It's important not to cut off all of the spears. If you did that, the plant would probably give up and die. But if you cut two or three and leave two or three, the plant is encouraged to keep growing. They grow so fast; you might be surprised. I harvest every other day – even every day in the middle of the season – because if they get too tall they will be tough. And we don't want that!

In the first year of harvesting, you should only cut for about four weeks as they need to build up strength. The season on the whole is about six weeks. Liz says eight, so we agree to disagree! I never pick after six weeks because I think you need to let the asparagus know that it's surviving. If you keep cutting things off, then it will give up, and it could affect the following year's crop. I believe it's important to let some grow naturally to help the plant go through the natural cycle of grow, harvest, flower and die. I always keep my bed weed free. When the asparagus has finished harvesting, I give the bed a sprinkling of general fertiliser. It pays to look after it, and treasure your asparagus plants, because it'll last you fifteen to twenty years.

## Harvesting in May

Asparagus

Lamb's lettuce

Lettuce

Radish

Rocket

Tomato

**Liz and I love June. We're blessed with long, hot days in the garden.** We enjoy sitting outside with a cold drink as the sun goes down after a hard day's work. Everything is growing very quickly (including the weeds!). There's lots to plant out, lots of seeds to sow, and lots of food to harvest. How could you not love June?

Even if you think of June as a warm month, don't get complacent about frosts! In 2020 there was a frost on 5 June – so keep one eye on the forecast.

If there isn't enough rain to keep the water butts full, I top them up from the tap and let the water stand for at least twenty-four hours to get rid of any chlorine.

Throughout the warmer months, I make sure to keep the atmosphere damp in the small greenhouse where I am growing tomatoes and cucumbers. We leave buckets of water in there to evaporate and spray the floor on hot days. A dry atmosphere encourages blossom end rot, which is mainly due to a lack of calcium. There's also a danger that you'll get red spider mite. These are tiny spider-like creatures that cause leaves to turn a mottled yellow, which hinders the plants' ability to photosynthesise, and the tomato and cucumbers will eventually die.

As well as the tomatoes in the greenhouse, I grow bush tomatoes outside. These are left to their own devices, I just

put a short cane by them to give support if they become top heavy. As soon as the first fruit on the bottom truss of the tomato forms, it's time to start feeding with a suitable feed. I think Tomorite is the best. I actually use this for feeding everything in the greenhouses, as well as some of the stuff growing outside.

In the large greenhouse, Liz will have cleared all her bedding plants so I can I move in, potting up sweet peppers, chillies and aubergines. I use 10-litre pots for them. I also start feeding these towards the end of June. They will stay in the greenhouse all summer.

Swedes and turnips can be sown now, as well as marrows or squashes if you haven't grown them in the greenhouse. Just remember that they need quite a lot of space. One of our favourite winter squashes is called Turk's Turban; they're very colourful and they have a lovely, sweet taste. I'm also trying a pale green variety called Crown Prince this year. June is the last chance to sow runner beans and French beans. If you sow any later they won't reach maturity in time to harvest. I sow a row of French beans at the allotment – I like to plant dwarf beans, not the climbing ones. It's now too late to plant maincrop carrots, but early carrots take a shorter time to mature, so I usually sow some of these now, to give us small carrots in the autumn.

If you have less space, in June you can sow some 'cut and come again' salad leaves in containers. I mix some different lettuce seeds with rocket and lamb's lettuce. I scatter the seeds reasonably thinly in a container. You can cut what you want when they're ready, and you can harvest several times before they run out of steam. Once they've given all they have, I remove the spent plants and replenish the container with some fresh compost and a bit of fish, blood and bone fertiliser, then start again. This ensures a continual supply of lettuces.

Everything from the greenhouse and cold frame that hasn't already gone out in late May is planted out this month. To protect my vegetables from birds, I make covers out of wooden battens, blue plastic water pipe, and fine insect mesh. They are quite light, so it's no problem to lift them up and move them around. I make sure all my brassicas are protected with these netting covers, because the pigeons love the young plants nearly as much as I do!

Flea beetle can become a pest, too. Especially on radishes. They eat little holes in the leaves. I drape some horticultural fleece or fine netting over the radishes to stop the beetle. And I keep up my slug patrol. The last thing I want is to let all my hard work go to waste!

## Planting out in June

Aubergine

Brussels sprouts

Cabbages

Cauliflower

Cavolo nero

Chard

Chilli

Courgette

French beans

Marrow

Onions

Pumpkins

Runner beans

Sprouting broccoli

Squash

Sweet pepper

Tomato

This month, I'll start harvesting veg I planted last autumn, such as beetroot, onions and spring cabbage.

June is also the time to harvest the garlic. There are two types of garlic: hard neck and soft neck. Hard neck generally has a bigger bulb, but it doesn't keep so well – that's what I think, anyway. You plant the hard necks in the autumn, and the soft necks in the spring. But they both get harvested at the same time. Once harvested, if the weather is lovely and sunny, I put the bulbs on a table or bench to dry off. If it looks like rain, I put them under shelter. If they stay wet, they will rot. I pickle garlic in vinegar. I like it on a pizza – it's lovely, pungent and sweet.

I will be digging up my first early potatoes this month. The first time I dig I will need two or three plants to give us a boiling of small potatoes. But as the days go by, the potatoes get bigger. We usually reach a point when one plant gives enough decent-sized potatoes to feed us for two or even three days. With the tubs of potatoes I have in the garden it's easy to gently feel around in the compost to find the ones that are big enough to harvest and leave the rest to carry on growing.

We will be eating lots of salads this month, so I try to keep up a continuous supply of lettuce and salad leaves, tomatoes, radishes and spring onions.

## Harvesting in June

Beetroot

Cabbage

Garlic

Kohl rabi

Lettuce

Onion

Potato

Radish

Spring onion

Tomato

**By July, we're well into summer. Everything is growing and cropping well at this time of year.** I could do with four arms and legs to keep up with it all!

Although the weather is usually lovely now, there is always a risk of winds so it pays to make sure everything is as secure as possible. A high wind can do a lot of damage with everything in full leaf. The runner beans in particular can get a hammering.

But with any luck the weather will be glorious, and the days will be long, but not too hot. During the summer I like to get up really early – about 4 a.m. – and get out while it's still cool. I'll work through till lunchtime and have an afternoon nap during the hottest part of the day. We have a little gazebo out in the garden and I'll quite happily have a snooze in there. Only for about an hour – I don't want to miss too much of the sunshine! After my nap, I'll work through till dinnertime.

It's the best feeling in the world to finish up a long summer's day, tired from a day's work. I feel like I'm glowing from all the fresh air and sunshine, and I love to pour a nice cool drink and, as they say, catch the last of the rays.

I spend my evenings watering the greenhouses, the garden and allotment when necessary. Even if it has been raining, my plants in containers outside don't get enough

moisture as the foliage is covering the surface by now. As I have so many containers, I use a hosepipe and push the nozzle under the leaves to ensure the water gets into the compost. It's always best to water in the evening as it gives the ground more chance to absorb the moisture. Also, if you're watering in the morning, there is the possibility that the plants will get scorched by the sun. A drop of water on a leaf will act like a magnifying glass.

There are a few odd jobs to keep on top of in July. My broad beans finish cropping this month so I cut them down to ground level and leave the roots in the ground. All beans and peas feed on nitrogen, and as they grow they form little nodules on their roots packed with this vital stuff. It's a brilliant natural fertiliser for your soil.

I keep sowing and planting lettuces. We have salads well into autumn, so the sowing doesn't stop! I like to sow some rocket in a shaded area because it doesn't like too much heat. The only trouble is that flea beetles also love rocket. If necessary, I will cover the rocket with fleece. It's amazing how useful fleece can be. I sometimes wonder, how did we ever manage before it?

If there is a spot of free ground on the allotment or at home, I will sow some green manure. This is a crop that is grown for the purpose of being cut and dug back into the ground. It is ideal for returning nourishment to the soil. There are lots of plants used for this – you can find different examples in any vegetable catalogue or garden centre.

It's time to plant out the winter cabbage, broccoli, Brussels sprouts, and kale that I have sown in rows in my seedbed. I water the row of young plants first to help loosen the roots, then gently pull them up, plant out at the correct distance, then cover them.

I think the biggest problem we have this month is with cabbage white butterflies. They're named for a reason – their caterpillars just love munching on your cabbages and any other brassicas you've got growing. Even with all the precautions I take to cover them up with netting, I quite often find a butterfly flitting about inside my tunnels. I'm convinced they get on the ground and look for any gap between the cage and the soil. I suppose I can't blame them – my cabbages are quite delicious! I regularly inspect them and pick off the caterpillars or the little clumps of yellow eggs, otherwise my entire cabbage patch could go to waste.

I am not a completely organic gardener, but if I can find a solution to a problem without using chemicals, I do. I think it's much better to go without the chemicals wherever possible. Sometimes that means spending a bit more time to manage the problem, but I don't mind too much.

### Planting out in July

Broccoli

Brussels sprouts

Cabbage

Cauliflower

Kale

Lettuce

Potato

This month, as well as continuing to baulk up my main-crop potatoes to keep the young tubers in the dark, I like to plant a few containers of potatoes. One of the benefits of growing potatoes in tubs is that they are pretty safe from keeled slugs; these are the little black things that eat holes in your potatoes. So, to avoid trouble with the little critters, it can be worth growing all your spuds in containers.

I try not to use slug pellets. Many things eat slugs, including blackbirds, and we have a good number of them in the garden. A slug that has ingested pellets doesn't do a blackbird any good if the bird eats it.

There's a lot of produce ready to harvest in July, such as summer cabbages and cauliflowers. When I see a small head of cauliflower showing, I gather up the outer leaves around it and use a rubber band to hold them in place. This keeps the cauli head nice and white as it grows to maturity.

This month, the shallots are ready to harvest. I just loosen them with my fork and then pull them up by hand. I leave them on the ground for twenty-four hours to start drying off. The next day I bring them home from the allotment and lay them out in the sun to finish. If rain is forecast, they go on the bottom tier of the greenhouse staging. I'll make sure to keep checking the shallots once a week to make sure none are going off.

I'll also be starting to harvest my second early potatoes. Of these, Charlotte is our favourite. There's nothing nicer than a plate of homegrown salad, a bit of cold meat, and some Charlotte potatoes on the side. We just wash them off and boil them with the skins on. Fabulous. I gave away a couple up the allotment a few years ago. The following year everyone was planting them!

We can get overrun with vegetables during July, so we start making chutneys and pickles. Liz loves her caramelised onion chutney. My favourite is piccalilli – Liz doesn't like it, though. Never mind – more for me! If we still have surplus after this (which we usually do) the family will get a share and I'll take some to one of the old people's homes.

Every Sunday morning local couple David and Joanna put the 'Milton Table' – named after our village, Milton-under-Wychwood – by their gate. Everyone in the village brings their spare vegetables, fruit, flowers, plants and even packets of seed and jars of pickles and jams. Then people come and take whatever they like and give a donation to a local charity, the Lawrence House Trust. I think it's a wonderful idea – every village should have a table.

Bartering on the allotment is also great at this time of year. I think it's good to share what we can. It would be a crying shame if the food went to waste!

## Our easy coleslaw

- Shredded white cabbage

- Finely chopped onion

- Shredded or grated carrot

- Mayonnaise or salad cream (or a mixture of both)

*Mix it all together and that's it!*

**Well, here we are in August. This is a month of harvest, harvest, harvest.** I'm not planting much. On the whole, August is a month that is about enjoying our rewards for all the hard work earlier in the year!

However, the middle of August is when I like to plant up some Charlotte potatoes in a tub to give us some new potatoes for Christmas day. Last year my tubs of potatoes were fantastic, I had just under 4 kg from two seed potatoes in a 30-litre container. If you like kohl rabi as much as we do there is still time to sow a few hardy purple ones to harvest in October. And you can sow chard for winter harvesting now too.

Just because I'm not so busy sowing and planting doesn't mean I'm taking it easy. For a start, August is show season, so I'll be harvesting and preparing all the big veg I've been hoping will beat my previous years' records.

The break from sowing and planting also presents a good opportunity to make repairs on any wooden features in the garden and on the allotment before I get busy again.

This year, I'm making a vermin fence. It's about a metre tall with a wire mesh, like chicken wire, and will hopefully stop animals such as badgers coming into the allotment. I had to get permission from the allotment association – it's important to make sure you have that before putting any

new fencing up. My allotment is on the side of a footpath, and we get an awful lot of dog walkers. Quite often dogs run over the allotment. I'm not one for falling out with people, so the fence will cure that problem as well!

Oh, and one last thing – don't forget to keep up with the weeding and, if it's dry, the watering!

In the greenhouse, the tomatoes, peppers, chillies and aubergines are growing well. I keep them well watered and fed twice a week. The aubergines need to have the growing tips taken out after five or six fruit have formed, in order to give them a chance to grow to a good size.

I'm very busy digging my second early spuds, cutting the Turk's Turban squashes, Patty Pan (beautiful little yellow squashes), courgettes, marrows, butternut squash . . . we love them all! I make sure I cut the courgettes when they are small as the more you cut the more you get.

The cucumbers will still be cropping well. One of the reasons I grow Mini Munch is because it seems to be more resistant to mildew on the leaves than a lot of cucumbers. So many times, my cucumbers have grown well for a couple of months but then have suddenly got mildew, turned up their toes, and died.

I always hope for August to be a lovely dry month, as that's ideal for lifting and drying the onions up on the

**Sowing in August**

Chard

Kohl rabi

Lettuce

**Planting in August**

Potato

allotment. They're ready to lift when the leaves start to flop down and turn yellow. I go along with my fork first, gently loosening them, just like with the shallots in July. Then I can pull them up with no worries about damaging them. If the forecast is good, I leave them for 24–48 hours before bringing them home, where they can finish drying off on the staging outside.

About the second week of August, we set about preparing and pickling the shallots. We make ten 2-litre jars full. Quite a lot of shallots! It takes about a week altogether. It's nice to sit out in the shade on a lovely day and skin them. This saves us from stinking out the house with the onion smell and you cry less in the fresh air.

When we've skinned them all, we wash and dry them and then soak them in salt for twelve hours to suck out all the moisture. Then we wash and dry them again. Then I fill half a jar up with shallots, add one tablespoon of sugar and a sprinkle of pickling spices, and fill it to the level of the shallots with vinegar. I then fill the jar up with shallots and pour more vinegar over very slowly. Then I place the lid on top, but I don't screw it up. Instead I leave it overnight to let all the air escape. In the morning, I top the jars up with some more vinegar, screw the lids on, label them, and put them in storage. We won't open them until Christmas Day.

We love our caulis.
Cauliflower cheese
— beautiful!

**I really think September is my favourite month of the year.** The searing heat of July and August has gone, and a lovely, pleasant warmth takes its place. It's a time to slow down just a little and really enjoy the weather and the garden. The early crops have all but finished, but there are still beans, courgettes, tomatoes, sweet peppers, chillis, aubergines, cucumbers and sweetcorn to enjoy.

At its best, September is a wonderful time. But if we are unlucky and it turns out to be a very wet month, September can be a challenge. We are still harvesting lots, which you don't want to do when the produce is wet, because it will rot.

As soon as I finish harvesting something, the foliage is cut down. Unless the plant is diseased, I'll put it on the compost heap to rot. With everything except for peas and beans, the roots are pulled up as well. If there's any risk of disease in your ground such as club root – a nasty disease of the cabbage family – don't put the roots on the compost. Put them in the bin. If you put them in garden waste, they might come back to haunt you.

Where the green manure has grown enough, I will chop it down and dig it back in. I'm old-fashioned enough to do a process called double digging. I don't do it every year, but I do a certain area each year and gradually move around my plot over a few years. I believe it's important because you

can get a 'pan' underneath where you normally dig. A pan is when the soil becomes compact and water can't drain away so readily, especially on clay soil. When you double dig, you break the pan up, allowing free drainage.

First, I dig a trench about 30 cm wide and one spade's depth across one end of the plot of land I'm going to double dig. I fill my wheelbarrow with what I have dug out and move it to the other end of the plot and dump it – you'll see why in a minute!

Then I dig the bottom layer of the trench with a fork, right down as deep as the fork will go. When I've done that, I dig the next trench, this time turning the soil and green manure over into the first trench that I made. I repeat this process across the whole plot until I get to the other end and finally fill that last trench with the soil and manure I dug out at the beginning. It's hard work, but extremely satisfying. And it keeps me fit!

I also now prepare my ground for my winter hardy onions that mature in the gap between running out of my stored onions and harvesting the maincrop next year. The variety I use is a Japanese one called Senshyu Yellow. After preparing the ground, I sprinkle some all-purpose fertiliser on it. I use a multipurpose growmore or Vitax Q4, then leave it to settle until next month.

There is just time to sow a last lot of salad leaves and lettuce in September, to take us through the autumn. With this last sowing I make sure to cover the plants with a cloche as soon as it gets cold.

A great job for a chilly morning is turning the compost. It's good exercise – it certainly keeps me warm. I like to have two compost bins on the allotment. Mine are made with old pallets. One will be the stuff from last year that has rotted down, and the other is full of stuff from this year. I will have been using the old compost throughout the summer so this bin will hopefully be empty by September. So I turn the heap of newer compost into the empty bin. Turning the heap now aerates it so it will generate more heat. The heat is caused by the microorganisms digesting the waste material and breaking it down into that lovely compost – they need the oxygen to live and work well.

This month I dig my maincrop potatoes up. I choose a time when the forecast is clear for at least a couple of days. Once they're all out of the ground, I leave them for at least four hours to let the skins firm up as they are easily skinned when they first come out but the drying process will help to preserve them through the winter. Usually, I spend the morning digging them up then I'll go home for lunch and a rest. Then I'll go back up the allotment and sort them out

**TIP**

When you're digging potatoes, always push the fork in from the side of the row and don't use unnecessary force. This way you're less likely to damage the potatoes as you lift them.

**Harvesting in September**

Cabbage

Celeriac

Chard

Leeks

Potatoes

Pumpkin

Swede

Winter squash

while I'm bagging them up. Because they're harvested when it's nice and dry, there shouldn't be much dirt on them.

I bag the large potatoes separately to keep solely for baking. We both love a baked spud. I'll have mine with tuna and sweetcorn; Liz prefers beans. Sometimes just a piece of good-quality butter melted on the potato with a pinch of salt is lovely. Last year I had ten potatoes that weighed nearly 6 kg altogether. Liz said she couldn't manage one of them, so I had to eat them all. What a hardship!

This is a quiet month for sowing and planting, but there are plenty of vegetables to harvest. Celeriac is lovely cut into chunks and roasted, leeks may be ready to eat, and some swedes are big enough to use.

If the winter squash have stopped growing, I will cut them from their parent plants, remove the old plants, and leave the squashes to cure in the sun for a few days before storing them in my frost-free garage. If rain is forecast, they go into the greenhouse to cure. Curing makes the skins ripen and go hard and helps to preserve the squashes. Some varieties are very good and can be stored really well all through the winter.

I usually put some pumpkins aside for the grandchildren and great-grandchildren for Halloween. But not enormous ones – they can't lift them!

**The days are getting shorter, the sun is lower in the sky, and the leaves are changing colour.** The temperature is dropping, especially at night, so any time now we could get the first hard frost.

I like to have a day off on the 24th as it's my birthday. Liz cooks something special for dinner and my daughter Karina will bake me a cake.

It's a good idea to cover the last lettuce and salad leaves with cloches before the frosts come. Frost will also kill off any outdoor tomatoes, so I cut off any remaining bunches and take them into the greenhouse to finish ripening off. I know a lot of people talk about what you can do with green tomatoes, but I find that virtually all of them will ripen if they're left on their vines and put on the staging of a cold greenhouse. We can be eating them in December.

Cucumbers have probably finished by now so I just clear away the plants and compost the whole thing. Some people have a shredder, but I just get a pair of secateurs and chop things up. The harder something is, the longer it takes to rot, so I chop up the woodier things into smaller pieces.

I also get all the marrows, courgettes and squashes under cover and compost the stalks. Any remaining courgettes need eating as they won't keep, but with any luck the rest will last a good part of the winter.

We'll always end up with a few peppers and aubergines outside because the greenhouse gets full and we can't bear to throw them out! If there any of these left by now, I bring them into the greenhouse to finish growing any fruit still on them. Then I compost them as well.

One job I like to do is up the allotment. I'm a bit old fashioned I suppose: I like to see my plot neat round the edges. So, using a garden line, I make a line down each side of the plot. Then I dig a nice straight edge, throwing the dug soil on to the bed, and leaving a furrow. The furrow stops weeds creeping from the paths surrounding the allotment on to the beds. It does look lovely and neat.

October is the one time of year when I will have a bonfire on the allotment. I don't like to burn things often, as the wind can take it and sometimes people don't like bonfires. I just have one to burn rubbish. Don't worry – only natural stuff! No plastic or rubber. The resulting ash I will use on the garden as potash.

This month is mainly given over to cleaning and tidying up the garden and allotment. October is a good time to clean water butts out so they can fill up with fresh rainwater through the autumn and winter. I take down the bean sticks, bundle them up and tie them with some thick string or thin rope and put under cover until next spring. And there are always ideas of what needs to be done buzzing around in my head. I'll be coming up with plans for winter.

One of the really good things about this month is that courtesy of the Milton-under-Wychwood Allotments and Gardens Association we get a catalogue from Kings Seeds. We have a great time choosing the seeds we want for the following year and planning where they will go. The best bit is, they're half price! You can't go wrong.

It's good practice to rotate crops to prevent diseases building up in the soil. Most gardeners have a three-year rotation. For example, once my brassicas are finished, I will then cultivate that plot and plant root crops, mainly potatoes. The only bed that stays the same year after year is my onion bed, as onions are one of the few things that you don't have to rotate.

As things are a bit less hectic, October's also the time to take the tomato supports down in the greenhouse and have a general tidy up. In the larger greenhouse, on a nice day with no frost forecast, Liz and I clear all the plants out. Over the previous month, dahlias, chrysanthemums and various other tender plants have been brought in for the winter. There always seems to be a crossover – the peppers and aubergines aren't finished when the flower plants need to come in. Next I clean the staging and the glass. Then I put

**TIP**
Please don't forget the weeding. It's so important. My dad used to say, 'Never walk past a weed.'

a couple of sulphur candles on the floor and light them. I get out of there sharpish, and make sure the door and windows are shut. Within twenty minutes, the whole greenhouse is full of lemon-coloured smoke. Lots of people now prefer to use garlic candles as you have to be very careful with sulphur. But the burning sulphur has the effect of getting rid of any pests lurking in the corners. It must be kept shut overnight. The next day I open the door and leave it to air for several hours. After that, I cover the inside of the greenhouse with bubble wrap for insulation and then we can get the plants back in for the winter.

It's another quiet month on the sowing and planting front, but if I haven't had time before, I plant my broad beans, garlic and Japanese onions.

My choice for beans is Aquadulce Claudia – a very good winter broad bean that never lets me down. They're lovely – I'd recommend them to anyone. I sow the beans in a double row. First, I draw a line across the garden. Then I take my dibber and dib a hole about 5 cm deep and pop a seed in. Then I go to the other side of the line, dib another hole, and plant another seed, zigzagging across. I put wire netting covers over them to protect against pests. I expect these to crop at the end of May, or beginning of June, but the joy of picking early broad beans when there isn't much else cannot be over estimated.

## Sowing in October

Broad beans

## Planting in October

Garlic

Japanese onion

Shallot

In October it's time to harvest the parsnips. I wait until after the first sharp frost, as this turns the starch to sugar and they taste so much better. I'll also be digging up beetroot and carrots and putting them away to store and eat through the winter. There is one thing I really need to get off my chest. If you are going to store any root crop, please *don't* wash them first! I keep preaching this. If the soil is left on the vegetables, they will last a lot longer than if they are washed clean. When I dig parsnips and carrots in the winter, I leave them in a trug outside the back door unwashed. They will last for days, if not weeks. As soon as you wash a vegetable, the rotting process begins. The deterioration starts immediately. Within a few days, they will be unfit to eat. Best to leave them protected by the soil, the way nature intended.

Towards the end of the month, the famous Milton Table is decorated with all colours of pumpkins and squashes. A couple of years ago, I suggested a pumpkin day and – without tooting my own horn too much – it worked really well! We raised £69 in two hours for the Lawrence House Trust. People can make a donation and take home a pumpkin to carve for Halloween.

On the 31st, I carve one of my own pumpkins and put it outside. Why should the kids have all the fun?

**November is the month that I have the least amount of work to do in the garden and up the allotment.** I don't mind – the weather can be pretty ropey, and it's a fine time for a bit of a rest after the busy summer season. Saying that, I am usually busy doing something. There's always a list of projects waiting to be completed!

But when the weather is foul outside, I like to curl up with a load of catalogues and dream about what else I can buy and grow. Around this time, I'll place an order for my potatoes from a reputable dealer. It's wise to order early in case anything is in short supply.

Once a month I check on the potatoes and onions in store. You can usually smell a rotten potato if there's one in a sack. I always store mine in proper paper potato sacks. They must be kept out of the light or they will go green and green potatoes can be poisonous.

During November, I take the opportunity to dig any vacant ground. We live on very sandy soil, so when I have dug the garden, I lay thick black polythene over the soil and peg it down. This helps to stop the nutrients draining out of the soil if the winter is wet. It also has the advantage of helping the soil to warm up earlier in the spring. On the whole, I am glad we live in an area with sandy soil. Sandy soil is very light and easy to cultivate. If I lived on clay soil,

I would approach it differently. I would dig the ground with a spade, turning the soil over in big lumps, and then leave it for the winter frosts to break down. This does it the world of good.

Once a year, I like to dig a load of manure into the allotment. I'm lucky enough to have a friend whose wife keeps horses. He kindly gives me a couple of tons.

In the smaller greenhouse, I will have my tubs of Christmas potatoes growing. I put a heater in there and set it just above freezing. I don't need actual heat in there, but I don't want the potatoes killed off by a hard frost.

It's time to tuck up the asparagus bed for winter. The ferns will have turned yellow by now and are better out of the way, so I cut them down and put them on the compost. Then I put a layer of manure or compost over the top and it's all done.

This month there are so many vegetables for us to eat: Brussels sprouts, winter cabbages, spinach, chard, carrots, parsnips and celeriac. Jerusalem artichokes, too. We enjoy them steamed or you can have them roasted, like potatoes. Don't eat too many though, else you'll be embarrassed. They can make you quite gassy, let's say!

November is also the time to lift any turnips you've got left as they aren't hardy enough to withstand the freezing weather. Swedes should also be ready to eat. If there is frost forecast, I nip up the allotment and dig up some root vegetables, bring them home and put them outside the back door so we don't go without. And yes, you've guessed it, I will store them unwashed! There are certainly plenty of vegetables to eat now. No complaints from me!

If you sowed a last row of lettuce under cloches, they should be growing well and will need thinning out. If you're lucky, you might get a meal out of the lettuces pulled up. Waste not, want not!

The last of the celery can be lifted, too. I found when I lifted mine that the outside stalks weren't great, but we had the celery hearts roasted with other vegetables with our Sunday joint and they were wonderful.

If I've collected my own seed from things like peas or beans, I like to test the germination of the seed before I store it for the winter. I take ten seeds and put them between wet kitchen towel sheets in a Tupperware container. Then I put the lid on loosely and put it in the airing cupboard. I check it every day until I see the seeds growing. I count how many out of ten. That gives me a percentage and I know if it's worth saving them or not. So far, I've never had a failure, but it's always worth checking. Imagine storing them all winter to find that they don't germinate!

## Harvesting in November

Brussels sprouts

Cabbage

Carrot

Celeriac

Celery

Chard

Jerusalem artichoke

Parsnip

Spinach

**Well, here we are. It's nearly Christmas and we have the New Year to look forward to.** The birds are hungry at this time of year, and sometimes, when I'm not very busy, I'll just get a fork and dig over a patch of ground for the robins and blackbirds. They seem to almost demand that I do this, following me around until I give up and do what they are asking. They can hardly wait for me to get going and it's not unusual for me to have a robin and a pair of blackbirds under my feet, pecking at anything edible they can find.

We buy apples for the blackbirds, and I'll put out loads of suet balls, and niger seed for the goldfinches. We have a couple of resident wood pigeons, like everybody has! One of ours Liz calls Fred. He sits on the little bird table outside our sitting-room window and eats out of the feeder that is conveniently hanging off it. Our robin has actually learned to perch on the feeder and eat, which is clever for a ground-feeding bird. Usually, you'll only see robins feed off the floor or a table.

This month is all about working around the weather. If it's fine, I like to finish jobs that are outstanding. I'll check the nets over the cabbages, top up the compost bins, check wires and ties on things, clean any tools, seed trays, and pots that I haven't yet done. It doesn't matter how thorough

I am, there's always something hiding somewhere, trying to avoid the clean up!

I also move everything in my Cave. Jet, our little black cat, joins me. The mice like to come inside and chew up all sorts of things to make their nests with but Jet will soon let me know if there are any about. He very rarely catches one, but the smell of a cat helps to deter any mice. I'll also use this time of year to service things like my lawnmower and my rotavators, so they are ready for when I need them next year.

The week before Christmas, Liz and I will make the trip to deliver wreaths to our respective families. While we're in the village of Cassington where I grew up, I go (with permission!) to collect some mistletoe from a tree that I squashed a berry into some fifty years ago. That is a wonderful feeling.

On Christmas Eve, I'll go up the allotment and exchange season's greetings with other allotmenteers. On the same mission, I collect Brussels, parsnips, carrots and salsify and bring them home, ready to go with our potatoes for Christmas dinner.

While I'm up the allotment, Liz cuts some holly from the bush near the back door. It's a superstition, but we put a sprig of holly in every room in the house from Christmas

Eve until Twelfth Night. It's an old tradition in Liz's family that's meant to bring good luck and keep bad spirits away. Liz says you can't be too careful.

When I get home from the allotment, we go to the greenhouse and tip out the potatoes growing in the tubs. It's a moment of great excitement to see how many we've got. Last year, we decided to make a short video of our 'detubbing' for Twitter. Liz had my iPhone to record, and as I tapped the tub out for them to fall on the ground, Liz didn't realise she'd made an excited 'woo!' sound in the background. It's an early Christmas present for us, that's for sure! There's enough for Christmas Day and Boxing Day, and my children and neighbour will normally get some, too.

Come the evening, it's my job to prepare the vegetables ready for the big day. I do so happily, while scoffing the odd homemade mince pie. The mincemeat is an old family recipe made by Liz's sister in Scotland and sent down to us. It's very moreish.

On Christmas morning, we pick some lettuce for a prawn cocktail starter. For dinner, we get the turkey or goose on to cook, along with the vegetables. Before Covid, we would be welcoming family members. Hopefully the future will be brighter. We never miss the Queen's speech. After an afternoon snooze, we'll enjoy some cold meats with pickles.

**Harvesting in December**

Brussels sprouts

Cabbage

Carrot

Chard

Parsnip

Potato

Salsify

Generally, we're so stuffed after the Christmas dinner that we'll have our pudding for supper! Then we'll have bubble and squeak on Boxing Day – that goes down well with cold meat and cheeses, and all those pickles and chutneys we've been storing.

The week between Christmas and New Year is taken up with eating all the leftovers. I always end up making a turkey curry at some point. Like most people, we always cook too much food.

We don't do so much for New Year's Eve. Liz will make a wholemeal loaf and I'll buy a good-quality butter. I love a Cornish butter with salt crystals. We'll enjoy a cosy night in, with our freshly baked bread and butter, some smoked salmon, and a little bit of lemon. Oh, and don't forget, a bottle of champagne. It's pretty simple, but we like it. After the hectic Christmas season, we don't want anything complicated that takes hours to prepare.

We like to spend some time taking stock of what we have achieved in the year, and what we are hoping to achieve in the next year. We always think we can do better. It's called ambition! But I think New Year's resolutions are made to be broken. You can make a resolution any time you like, can't you? We just toast to all our family and friends.

Health and happiness – cheers!

## Acknowledgements

Well, that brings us to the end of my little book about big veg. It has been a real pleasure to write, and I hope that the information provided inside will be of use for people.

There are a few people I would like to thank for their part in helping me to write.

First of all, I'd like to say an extra special thank you to my partner Elizabeth. She kept me on task throughout the writing process, even when I was trying to relax and watch TV! I am very grateful for her hard work and guidance throughout the process, and for her support at every step along the way of this new journey we are on.

To all my wonderful followers on Twitter, I'd like to say a massive thank you for supporting me and my work in the garden. Without you, none of this would have been possible.

I'd like to send my thanks to every seed-growing company I've ever bought from over the years. Without them, I'd have nothing to grow!

I'd also like to note my gratitude to Philip Bowne for helping me with the writing of this book.

Last, but certainly not least, I want to say a massive thank you to Fiona Crosby and the rest of the team at Headline for providing me with the opportunity to publish this work.

# Disclaimer

Every effort has been made to ensure that the information contained in this book is accurate, but represent the author's personal experiences rather than expert guidance. Neither the publisher nor the author can accept any responsibility for any loss or damage allegedly arising from the use or misuse of any information, advice or suggestion in this book. The naming of any product or company in this book does not imply endorsement by the publisher or author and the omission of any such names does not indicate disapproval.

First published in 2021 by
HEADLINE PUBLISHING GROUP

1

*Photo credits*

Pages 4, 7, 8, 15, 18, 21, 24, 28, 31, 43, 53, 57, 65, 74, 79, 84, 86, 92, 104, 107, 136, 139, 144, 147, 151, 162, 191 © Sophie Davidson

Pages 2, 35, 39, 47, 49, 52, 55, 59, 62, 67, 70, 77, 81, 87, 89, 95, 97, 99, 101, 109, 112, 115, 117, 118, 123, 126, 128, 131, 153, 157, 158, 161, 165, 167, 171, 173, 176, 179, 181, 185, 188 © Gerald Stratford/Elizabeth Stratford

Cataloguing in Publication Data is available from the British Library

Hardback ISBN 978 1 4722 8701 4

Designed and typeset by EM&EN
Colour reproduction by AltaImage
Printed and bound in Italy by LEGO S.p.A.

HEADLINE PUBLISHING GROUP
An Hachette UK Company
Carmelite House
50 Victoria Embankment
London EC4Y 0DZ

www.headline.co.uk
www.hachette.co.uk